Uncovering Mathematics

with Manipulatives and Calculators

Levels 2 & 3

Developed by
Jane F. Schielack and Dinah Chancellor

Design by
Reecie Ross

With contributions by
Yolanda Andrade, Bob Fedorisko, Gay Riley-Pfund,
Jan Stevens, Lynn Tanner, and Dianna Tidwell

Acknowledgments

Texas Instruments would like to express appreciation for the contributions of several teachers, who helped with the evaluation of these materials.

Jacksonville, Texas ISD:

Lynn Dickerson, Selena Earle, Kathleen Glidewell, Frieda Grimes, Sandy Harrison, Debbie Jones, Donna McCown, Becky Nelms, Terry Robbins, Doris Ross, Susan Sekula, Kay Steelman, Cathy Stephens, and Susan Ward

Bryan, Texas ISD:

Paula Burkhalter, Ginger Freeze, Pattie Holliday, and Brenda Marino

College Station, Texas ISD:

Leslie McGinnis and Cindy Wolfe

K-6 Dallas Summer Institute Participants:

Laura Bennetti, Yonkers, NY; Charlene Brooks, Parkville, MD; Teresa Chaney, Fort Worth, TX; Cathy Cromar, Sandy, UT; Jerry Cummins, Western Springs, IL; Leo Edwards, Fayetteville, NC; Sheila Evans, Baltimore, MD; Elizabeth Garnett, Chicago Heights, IL; Bonnie Hagelberger, Mahtomedi, MN; Katherine Hebert, Philadelphia, PA; Verdrey Madzimoyo, Newton, MA; Janet Niino, Fresno, CA; David Pagni, Callabasas, CA; Maria Scognamiglio, Brooklyn, NY; Mary Jane Tappen, Jacksonville, FL; Elizabeth Travieso, Pembroke Pines, FL; Jacquelyn Wade, Chicago, IL; Michele Weiner, Miami, FL; Ed Zegray, Montreal, Quebec

Important Notice Regarding Book Materials

About the Authors

Jane F. Schielack is an Associate Professor of Mathematics Education in the Department of Mathematics at Texas A&M University. As a former elementary teacher, she is interested in how teachers and students can use calculators to enhance mathematics learning in the elementary grades. Her past work in mathematics education includes participation on the writing committee of the National Council of Teachers of Mathematics *Professional Standards for Teaching Mathematics* and development of state inservice modules in mathematics for Prekindergarten through Grade 2. She is currently teaching mathematics to preservice elementary teachers and directing ongoing inservice to support classroom teachers in their quest for more effective mathematics instruction.

Dinah Chancellor is the Coordinator of Mathematics and Science for the Grapevine-Colleyville Independent School District in Texas. She has been involved in teaching children to use reasoning and creativity to solve problems for most of her professional life. Her interest in teaching students to reason mathematically has taken her into classrooms at all elementary grade levels. She has taught special education, "regular" education, and gifted education. She has also served as coordinator of mathematics, science, and gifted programs in Texas. As a contributor to *Arithmetic Teacher*, she has written several articles, including "Higher-Order Thinking Skills — A 'Basic' Skill for Everyone," Focus Issue, February, 1991. She has also written the "Calendar of Mathematics Activities" for *Arithmetic Teacher* for three years.

Calculators in Elementary Mathematics

Twenty years ago, one educational question being asked about technology was, "**Should** calculators be used in mathematics in the elementary grades?" The availability of technology in today's world has moved us forward to the question, "**How** can calculators best be used in teaching and learning elementary mathematics?"

The activities in this set of books are examples of situations in which students develop healthy attitudes toward the use of technology, not as a replacement for their mathematical thinking, but as an extension of their mathematical power.

The Level 1 activities (available separately), designed for primary grades, build connections between physical representations and the mathematical symbolism of the calculator.

The Level 2 & 3 activities, designed for intermediate grades, use the calculator as a data-generating device to support exploration of mathematical ideas.

With the calculator, students can explore a problem situation by producing a fairly large set of data in a relatively short period of time. By controlling variables in the problem while collecting the data and organizing the collected data in various ways, students can look for patterns and make conjectures. These patterns and conjectures lead not only to solutions for the given problem, but also to general understanding of important mathematical concepts.

Components of the Activities

Each activity consists of *Teacher Pages* that give directions for the activity and a student *Recording Sheet* that provides a method for organizing collected data.

The *Teacher Pages* for each activity contain:

- The content strand with which the activity is most closely associated: Number Sense; Patterns, Relations, and Functions; Measurement and Geometry; or Probability and Statistics.

- An icon indicating the suggested level of the activity:

  🔘 Level 1 Activity

  🔘 Level 2 Activity

  🔘 Level 3 Activity

  Level 1 activities can be found in *Uncovering Mathematics with Manipulatives and Calculators Level 1.*

- An **Overview** of the mathematical purpose of the activity.

- A list of related **Mathematical Concepts** addressed in the activity.

- An **Introduction** section, which includes problems that are posed in motivating, real-life settings and often contain literature connections.

- A section for **Collecting and Organizing Data** pertinent to the problem, including questions that require students to think carefully about any models, symbols, or procedures that they have chosen to use.

- A section on **Analyzing Data and Drawing Conclusions**, including questions that require students to look carefully for patterns in the data and translate those patterns into conjectures that can be investigated with examples or verified with mathematical reasoning.

- A concluding section on **Continuing the Investigation** with ideas for related problems and activities.

The student *Recording Sheet* for each activity is designed to provide a slightly organized, but open-ended, structure in which students can successfully record and analyze the data that they collect in the activity. The *Recording Sheets* are also intended to be used by students as models for organizing data in problems encountered outside these activities.

Goals for Calculator Use in a Mathematics Activity

The design for this collection of activities for integrating the calculator into elementary mathematics instruction is based upon three critical goals.

Goal 1: The activities will address the major components of the elementary mathematics curriculum.

The current recommendations for the elementary mathematics curriculum put forward by the National Council of Teachers of Mathematics states that elementary mathematics instruction should emphasize *problem solving*, *reasoning*, *communication*, and *connections* while engaging students in learning mathematical concepts.

The activities in this collection are clustered into four content strands:

- Number Sense

- Patterns, Relations, and Functions

- Measurement and Geometry

- Probability and Statistics

These four strands are woven together with the integrated use of operations and graphing. Each activity is based on a problem — sometimes real-life, sometimes mathematical, in origin. Each activity is developed around questions that encourage students to analyze data and make inferences, and every activity requires some kind of mathematical communication, either spoken or written.

There are no activities whose *main* goal is to teach a student how to use a particular feature of the calculator. For more information on the specific techniques of using the calculator, refer to the instructional materials for each of the calculators: TI-108, the Math Mate™, and the Math Explorer™. You may order the materials by calling **1-800-TI-CARES.**

Goals for Calculator Use in a Mathematics Activity (continued)

Goal 2: The calculator will be an instructional component of activities in which students are engaged in developing understanding of important mathematical ideas.

The use of a calculator in Kindergarten through Grade 6 mathematics is often thought of as a "reward" for completing paper-and-pencil computation or as "support" for a student having difficulty with paper-and-pencil computation. The use of the calculator in these activities, however, is only one component of an instructional plan for teaching and learning mathematics.

In these activities, the calculator is used mainly for generating data, with students deciding what data is needed and what processes should be used for generating it. Students must make decisions about what should and what should not be changed in the problem situation as the data is generated and how the data should be organized in order to discover the patterns that lead to important mathematical ideas.

On the left side of the *Teacher Pages* are questions that address the content of the activity and may or may not pertain to the use of the calculator. Along the right side of the *Teacher Pages*, marked with the calculator icon are questions that particularly address the role of the calculator in the activity.

Goal 3: The calculator will be used in an environment that encourages reasoning and communication.

Most of the activities in this collection are designed for students working together in small groups of two, three, or four. To encourage interaction within groups, the students in each group can work together using one copy of the *Recording Sheet*. Each member of a group can take on a particular role in collecting or organizing the data, and these roles can be rotated so that each member is involved in each aspect of the problem. With designated roles in solving the problem, a group can work with a single calculator, rotating its use among the members.

The purpose of the calculator in most of the activities is to generate questions rather than answers. The solutions to the problems and the conjectures about abstract mathematical ideas develop from students and teachers analyzing the data and discussing the questions that are generated in the activity.

Table of Contents

**Levels
2 & 3**

Levels 2 & 3

Activities:

1   100 or Bust

2   Random Remainders

3   Recurring Remainders

4   Remainder Rules

5   Names for One-Half

6   Patterns in Counting
       with Decimals

7   Names for 100

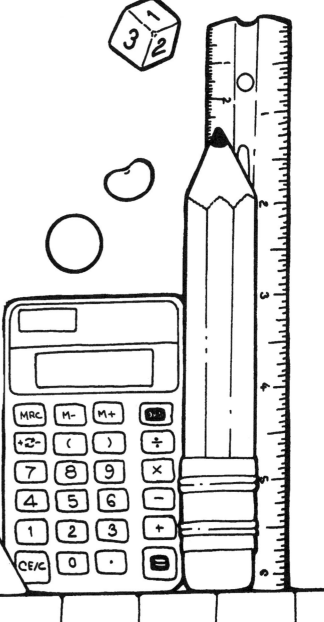

100 or Bust

Math Concepts

- whole numbers
- place value
- comparing numbers
- estimation
- addition
- subtraction

Materials

- TI-108, Math Mate™, Math Explorer™
- place-value materials
- **100 or Bust** recording sheets
- pencils
- number cubes (or dice)

Overview

Students will use estimation, place-value materials, and the calculator. They will place each of seven randomly generated digits on a place-value chart in either the ones or tens column to make a sum as close to 100 as possible without going over 100.

Introduction

1. Set up the activity by telling students: Seven people have a total of exactly $100. Each person has either all $1 bills or all $10 bills. How much money could each person have?

   Note: There are several possible answers.

2. Have students work in groups of three. While playing the game, the group will roll the number cube seven times. The result of each roll will represent either how many $1 bills or $10 bills someone in the group of seven people has.

3. On the overhead projector, model the game several times by rolling the number cube, demonstrating the three procedures/responsibilities, and explaining the rules.

   ### Procedures/Responsibilities

   1) The first student rolls the number cube. Based on the whole group's decision, he or she then records each of the resulting digits in the proper column on the place-value chart on the recording sheet.

   2) The second student uses place-value materials to represent the amounts on the hundred grid as they are written on the chart.

   3) The third student uses the calculator to keep a total by adding the amount of each roll of the number cube.

   ### Rules

   ▶ Roll exactly seven times.
   ▶ Place (write) each digit rolled in either the ones or tens place (column) to make a sum ≤ 100.

100 o la bancarrota

Conceptos matemáticos

- números enteros
- estimación
- valor de posición
- suma
- comparación
- resta

Materiales

- TI-108, Math Mate™, Math Explorer™
- materiales de valor de posición
- hojas de registro de **100 o la bancarrota**
- lápices
- cubos con números (o dados)

Resumen

Los alumnos utilizarán la estimación, los materiales de valor de posición y la calculadora. Colocarán cada uno de los siete dígitos generados al azar en una tabla de valores de posición en la columna de las unidades o de las decenas, para que la suma se aproxime lo más posible a 100, sin sobrepasarse.

Introducción

1. Organice la actividad, indicando a los alumnos lo siguiente: siete personas tienen un total exacto de $100. Cada persona tiene sólo billetes de $1 o de $10. ¿Cuánto dinero puede tener cada persona?

   Nota: hay varias respuestas posibles.

2. Que los alumnos trabajen en grupos de tres. En el juego, el grupo tirará los cubos con números siete veces. El resultado de cada turno representará cuántos billetes de $1 o de $10 tiene alguien dentro del grupo de siete personas.

3. En el proyector de transparencias, represente el juego varias veces tirando el cubo con números, demostrando los tres procedimientos/responsabilidades y explicando las reglas.

   ### Procedimientos/responsabilidades

   1) El primer alumno tira el cubo con números y anota cada uno de los dígitos resultantes en la columna apropiada de la tabla de valores de posición que viene en la hoja de registro, según la decisión del grupo.

   2) El segundo alumno usa los materiales de valor de posición para representar las cantidades en la cuadrícula de centena según aparecen en la tabla.

   3) El tercer alumno usa la calculadora para mantener un total, sumando la cantidad que se obtiene cada vez que se tira el cubo.

   ### Reglas

   ▶ Tirar exactamente siete veces.

   ▶ Poner (escribir) cada dígito resultante en cada turno en la posición (columna) de las unidades o las decenas, hasta llegar a una suma ≤ 100.

100 or Bust (continued)

Introduction (continued)

Example: The first student rolls a 2. The group decides to put the 2 in the "Tens" column. The first student writes 2 in the "Tens" column and 0 in the "Ones" column. The second student uses place-value materials to represent 20 on the hundred chart. The third student enters 20 in the calculator.

Tens	Ones
2	0

4. Have students play the game at least three times, rotating the responsibilities each time, so that each student gets to work with each representation. Each time they play, students should look for strategies to play a better game.

Collecting and Organizing Data

While students play the game, ask questions such as:

- How did you decide to place this digit in the ones place? The tens place?

- How does your sum affect your strategy as you play?

- What if we changed the rules so that you could go over 100, or so that you could choose to either add or subtract the number that comes up on the number cube? Could you get closer to 100?

▦ What does the recording sheet keep track of for you that the calculator doesn't?

▦ What do the place-value materials show that the calculator and recording sheet don't?

▦ What does the calculator help you do?

Analyzing Data and Drawing Conclusions

After students have played three games and recorded their data, have them work as a group to analyze the games. Ask questions such as:

- How did your strategies change within a game? How did your strategies change as you played more games?

▦ How did you use the calculator to help you decide what to do next in the game?

100 o la bancarrota (continuación)

Introducción (continuación)

Ejemplo: el primer alumno saca un 2. El grupo decide poner el 2 en la columna de las "Decenas". El primer alumno escribe 2 en la columna de las "Decenas" y 0 en la de las "Unidades". El segundo alumno usa los materiales de valor de posición para representar 20 en la cuadrícula de centena. El tercer alumno ingresa 20 en la calculadora.

Decenas	Unidades
2	0

4. Haga que los alumnos jueguen al menos tres veces, alternando las responsabilidades cada vez, para que cada alumno trabaje con todas las representaciones. Cada vez que jueguen, los alumnos deben buscar estrategias para jugar mejor.

Cómo reunir y organizar los datos

Mientras los alumnos juegan, haga las siguientes preguntas:

- ¿Cómo decidieron poner este dígito en la posición de las unidades o de las decenas?

- ¿Qué efecto tiene la suma en su estrategia a medida que juegan?

- ¿Qué sucedería si cambiaramos las reglas y pudieran pasar de 100 o escoger entre sumar o restar el número que sale en el cubo con números? ¿Podrían llegar más cerca a 100?

Cómo analizar los datos y sacar conclusiones

Después de que los alumnos jueguen tres juegos y registren sus datos, pídales que analicen los juegos como un solo grupo, usando las preguntas siguientes:

- ¿Cómo cambiaros sus estrategias durante un juego? ¿Cómo cambiaron sus estrategias a medida que jugaban más juegos?

🖩 ¿De qué mantiene un seguimiento la hoja de registro que la calculadora no hace?

🖩 ¿Qué muestran los materiales de valor de posición que la calculadora y la hoja de registro no hacen?

🖩 ¿Qué les ayuda a hacer la calculadora?

🖩 ¿Cómo utilizaron la calculadora para decidir qué hacer a continuación en el juego?

100 or Bust (continued)

Analyzing Data and Drawing Conclusions (continued)

- What if you did not have to roll exactly seven times? What if you could roll fewer times? What if you could roll more than seven times? How would your strategies change?

- Is there any game that you played that could have made a sum of 100 if you rearranged the digits? Use your recording sheet and calculator to find out.

Continuing the Investigation

Have students:

- Play the game with polyhedral dice other than cubes and see if their strategies need to change.

- Find a set of seven rolls that would equal exactly 100.

- Investigate how many sets of seven rolls they can find.

- Revise the game to include the 100s place and try to make a sum of 1000.

100 o la bancarrota (continuación)

Cómo analizar los datos y sacar conclusiones (continuación)

- ¿Qué sucedería si no tuvieran que tirar los cubos exactamente siete veces? ¿Qué pasaría si pudieran tirar más o menos veces? ¿Cómo cambiarían sus estrategias?

- En sus juegos, ¿hay alguno que pudiera haber llegado a una suma de 100, si reordenaran los dígitos? Para saberlo, utilicen su hoja de registro y la calculadora.

Cómo continuar la investigación

Los alumnos deben:

- Jugar con dados poliédricos que no sean cubos y ver si sus estrategias necesitan algún cambio.

- Encontrar una combinación de siete tiradas que dé exactamente 100 como resultado.

- Investigar cuántas combinaciones de siete tiradas pueden encontrar.

- Revisar el juego para incluir la posición de las centenas e intentar llegar a una suma de 1000.

Name:

100 or Bust
Recording Sheet

Collecting and Organizing Data

Game 1		Game 2		Game 3	
Tens	Ones	Tens	Ones	Tens	Ones

Hundred Grid

Strategies we used while we were doing this activity:

Nombre:

100 o la bancarrota
Hoja de registro

Cómo reunir y organizar los datos

Juego 1		Juego 2		Juego 3	
Decenas	Unidades	Decenas	Unidades	Decenas	Unidades

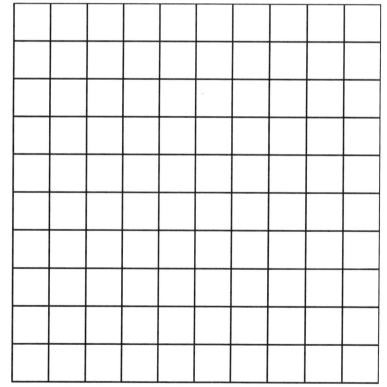

Cuadrícula de centena

Estrategias usadas durante esta actividad:

Random Remainders

Math Concepts

- whole numbers
- division
- graphing

Materials

- Math Explorer™
- **Random Remainders** recording sheets
- pencils

Overview

Students will use the calculator to investigate the relationship between divisors and remainders in whole number division.

Introduction

1. Present the following scene: You buy a bag of candy and want to eat it evenly throughout the week. What operation would you use to represent this on the calculator? (Division) Suppose there are 10 pieces of candy in your bag. How many pieces would you eat each day? How many would be left?

2. Have students enter **10** INT÷ **7** = into their calculators and compare the display to their mental computations. Discuss the meaning of each part of the display.

   Note: The quotient **1** represents how many pieces of candy there are for each of the seven days, and the remainder **3** represents how many pieces are left over.

3. Have students pretend that they could start with any number of pieces of candy at the beginning of the week. Ask students: If you divide the candy evenly over the week, how many pieces might you have left at the end of the week ?

4. Working in small groups, have students use INT÷ **7** = with a variety of numbers to collect data about the kinds of remainders that occur.

5. Have students record their remainders and keep a tally of each occurrence of a remainder on the frequency table provided on the recording sheet. Then have them organize their data and make a bar graph of their group's results on the grid also provided on the recording sheet.

Restos al azar

Conceptos matemáticos

- números enteros
- división
- representación gráfica

Materiales

- Math Explorer™
- hojas de registro de **Restos al azar**
- lápices

Resumen

Los alumnos utilizarán la calculadora para investigar la relación entre los divisores y los restos en la división de números enteros.

Introducción

1. Presente la situación siguiente: ustedes compran una bolsa de caramelos y desean comérselos equitativamente durante la semana. ¿Qué operación usarían para representar esta situación en la calculadora? (División) Supongan que hay 10 caramelos en la bolsa. ¿Cuántos se comerían cada día? ¿Cuántos sobrarían?

2. Haga que los alumnos ingresen **10** [INT÷] **7** [=] en la calculadora y comparen el resultado con sus cálculos mentales. Analice el significado de cada parte de lo que se ve en el visor.

    Nota: el cociente **1** representa la cantidad de caramelos que hay para cada día de la semana, y el resto **3** es la cantidad de caramelos que sobran.

3. Que los alumnos se pongan en la situación en la que pueden comenzar con cualquier cantidad de caramelos al principio de la semana. Después, pregúnteles lo siguiente: ¿cuántos caramelos creen ustedes que habrán sobrado posiblemente al final de la semana, si los reparten equitativamente durante la semana?

4. Haga que los alumnos trabajen en grupos pequeños y usen [INT÷] **7** [=] con varios números para reunir datos acerca de los tipos de restos que se pueden dar.

5. Haga que los alumnos anoten los restos y lleven la cuenta de cada vez que aparece un resto en el cuadro de frecuencias que viene en la hoja de registro. Luego, haga que organicen sus datos y creen un gráfico de barras con los resultados de su grupo en la cuadrícula incluida también en la hoja de registro.

Random Remainders (continued)

Collecting and Organizing Data

While students generate data about the remainders that occur when dividing by seven, ask questions such as:

- Do you think those are all of the possible remainders? Why or why not?

- What if you start with a really large number?

- How will you organize your data about the remainders you generated?

- How will you show your results on the graph?

- How do you think your results will compare with the results of other groups?

What do the two numbers generated by using INT÷ represent?

Analyzing Data and Drawing Conclusions

After students have made graphs of their data for their small groups, have them combine their data into a whole class graph and analyze the data. Ask questions such as:

- What remainders occurred when you divided by seven?

- How is your group's data like that of other groups? How is it different?

- Did the size of the number with which you started seem to matter? Why or why not?

- Do you think these are all the possible remainders? Why or why not?

- Did any remainders occur significantly more than others? Why or why not?

- What if you decided not to eat any candy on Sunday and spread it evenly over only six days? How do you think your results would change?

- How is the class graph like (different from) your group's graph?

How is the display generated by using INT÷ different from the display generated by using the regular ÷ key? How are they alike?

Restos al azar (continuación)

Cómo reunir y organizar los datos

Mientras los alumnos generan los datos acerca de los restos que se producen al dividir por siete, haga las preguntas siguientes:

- ¿Piensan ustedes que ésos son todos los restos posibles? ¿Por qué?

- ¿Que sucedería si empezaran con un número realmente grande?

- ¿Cómo organizarán los datos acerca de los restos que ustedes generaron?

- ¿Cómo representarán los resultados en el gráfico?

- ¿Cómo creen que sus resultados se comparan con los de los demás grupos?

 ¿Qué representan los dos números generados al usar INT÷ ?

Cómo analizar los datos y sacar conclusiones

Después de que los alumnos hagan los gráficos con los datos de su grupo, pídales que combinen sus datos en un gráfico general y analicen los datos. Haga las preguntas siguientes:

- ¿Qué restos se registran al dividir por siete?

- ¿En qué se parecen los datos de su grupo con los de los otros grupos? ¿En qué se diferencian?

- ¿Era importante el número con que empezaron? ¿Por qué?

- ¿Piensan que éstos son todos los restos posibles? ¿Por qué?

- ¿Algún resto se dio con mucho mayor frecuencia que otros? ¿Por qué?

- ¿Qué sucedería si decidieran no comer ningún caramelo el domingo y repartirlos equitativamente en solo seis días? ¿En qué forma piensan que cambiarían los resultados?

- ¿En qué se parece (y diferencia) el gráfico general con el de su grupo?

 ¿En qué se diferencia el resultado generado al usar INT÷ del generado al usar la tecla común ÷? ¿En qué se parecen?

Random Remainders (continued)

Continuing the Investigation

Have students:

- Investigate the remainders generated by three other divisors. Make a general statement about the relationship between divisors and possible remainders.

- Make a table of dividends and divisors that generate remainders of 0. Make a generalization about the conditions that generate a remainder of 0.

Restos al azar (continuación)

Cómo continuar la investigación

Los alumnos deben:

- Investigar los restos generados por otros tres divisores. Formular una afirmación general sobre la relación entre los divisores y los restos posibles.

- Hacer un cuadro de los dividendos y los divisores que generan restos de 0. Formular una afirmación general sobre las condiciones que generan un resto de 0.

Name:

Random Remainders
Recording Sheet

Collecting and Organizing Data

Frequency Table for Dividing by_____

Remainders										
Frequency										

Graph: How Often Each Remainder Occurred When We Divided by _____

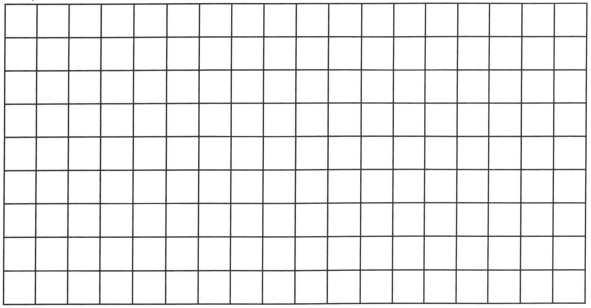

Conclusions we made about remainders from the data in our graph:

Restos al azar

Hoja de registro

Cómo reunir y organizar los datos

Cuadro de frecuencia en la división por _____

Restos										
Frecuencia										

Gráfico: frecuencia de cada resto al dividir por _____

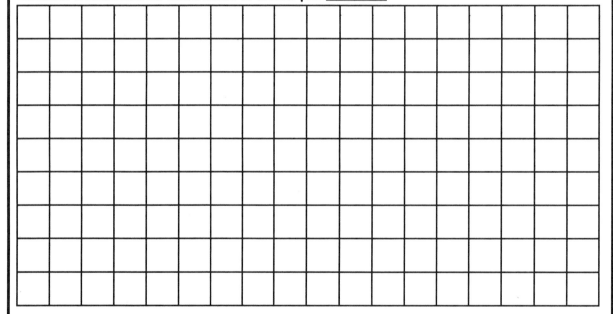

Conclusiones sobre los restos a las que llegamos a partir de los datos de nuestro gráfico:

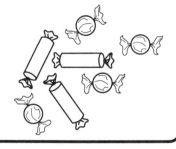

Recurring Remainders

Math Concepts	Overview
• whole numbers • division **Materials** • Math Explorer™ • **Recurring Remainders** recording sheets • crayons and markers • pencils	Students will use the calculator to investigate the patterns formed by remainders in whole-number division.

Introduction

> The **Random Remainders** activity on page 6 should be completed before beginning this activity.

1. Have students review the possible remainders that can occur when dividing by seven.

2. Ask students: Does any one remainder occur more often than any other? Explain your reasoning.

3. On the recording sheet, have students make a key by choosing a different color to identify each possible remainder, including zero.

4. Have students use INT÷ to divide each number on the hundred chart (see the recording sheet) by seven. To record the remainders, have students color the square on the hundred chart the color that represents its remainder on the key.

   Example: 19 INT÷ **7** = **Q 2, R 5.** If the color that represents the remainder 5 on the key is blue, students should color square 19 blue.

   Now, have students look for patterns.

Collecting and Organizing Data

While students are generating and recording their data, ask questions such as:

• What remainders occur when you divide by seven?

• What patterns do you notice on the hundred chart?

 How are the displayed quotient and the remainder related to the original number (the dividend)?

Restos recurrentes

Conceptos matemáticos

- números enteros
- división

Materiales

- Math Explorer™
- hojas de registro de **Restos recurrentes**
- lápices de cera y marcadores de fibra
- lápices

Resumen

Los alumnos utilizarán la calculadora para investigar los patrones formados por los restos en la división de números enteros.

Introducción

> Antes de comenzar con esta actividad, se debe completar el ejercicio de **Restos al azar** de la página 6.

1. Haga que los alumnos revisen los restos posibles que se pudieran dar al dividir por siete.

2. Pregunte a los alumnos: ¿algún resto se produce con más frecuencia que los demás? Expliquen por qué.

3. En la hoja de registro, haga que los alumnos elaboren una llave, escogiendo un color diferente para identificar cada resto posible, incluido el cero.

4. Haga que los alumnos usen [INT÷] para dividir cada número en la cuadrícula de centena (vea la hoja de registro) por siete. Para registrar los restos, haga que los alumnos coloreen el cuadrado de la cuadrícula de centena con el color que representa su resto en la llave.

   Ejemplo: 19 [INT÷] **7 = Q 2, R 5**. Si el color que representa el resto 5 en la llave es azul, los alumnos deben colorear de azul el cuadrado 19.

   En seguida, haga que los alumnos busquen patrones.

Cómo reunir y organizar los datos

Mientras los alumnos generan y registran sus datos, haga las preguntas siguientes:

- ¿Qué restos se dan al dividir por siete?

- ¿Qué patrones observaron en la cuadrícula de centena?

 ¿Cómo se relacionan el cociente desplegado y el resto con el número original (dividendo)?

Recurring Remainders (continued)

Collecting and Organizing Data (continued)

- What do these patterns tell you about the remainders that are occurring?

- What would happen if you divided by four instead of seven? By ten? How would your data look? Try it.

Analyzing Data and Drawing Conclusions

After students have explored with several divisors, have them analyze the patterns that they have seen. Ask questions such as:

- What patterns did you find in your data?

- Why do the colors go in a particular sequence on the hundred chart?

- What happened to the remainder when the dividend increased by one (and the divisor remained the same)?

- What happened when you used a different divisor? How did the patterns change? How did they stay the same?

- What do you think would happen if you continued beyond 100? Would the patterns continue? Why or why not?

- What color would 1000 be?

- What general statements could you make about remainders in whole-number division?

Continuing the Investigation

Ask students to describe situations where it would be useful to know the patterns in remainders in whole-number division.

How did you use the calculator to help you collect data in this activity?

How is the INT÷ key different from the ÷ key?

Restos recurrentes (continuación)

Cómo reunir y organizar los datos (continuación)

- ¿Qué indican estos patrones con respecto a los restos que se están dando?

- ¿Qué sucedería si dividieran por cuatro en vez de siete? ¿Y por diez? ¿Cómo se verían los datos? Inténtenlo.

- ¿Cómo usaron la calculadora para reunir los datos en esta actividad?

- ¿En qué se diferencia la tecla [INT÷] de la tecla [÷]?

Cómo analizar los datos y sacar conclusiones

Después de que los alumnos prueben con varios divisores, pídales que analicen los patrones observados, usando las preguntas siguientes:

- ¿Qué patrones descubrieron en sus datos?

- ¿Por qué los colores aparecen en una secuencia determinada en la cuadrícula de centena?

- ¿Qué sucedió con el resto cuando el dividendo se incrementó por uno (y el divisor permaneció igual)?

- ¿Qué sucedió cuando usaron un divisor distinto? ¿En qué medida los patrones cambiaron o permanecieron invariables?

- ¿Qué piensan que sucedería si continuaran más allá de 100? ¿Seguirían los patrones? ¿Por qué?

- ¿Qué color tendría el 1000?

- ¿Qué afirmaciones generales podrían formular acerca de los restos en la división de números enteros?

Cómo continuar la investigación

Pida a los alumnos que describan situaciones en las que sería útil conocer los patrones de los restos en la división de números enteros.

Name: _____

Recurring Remainders
Recording Sheet

Collecting and Organizing Data

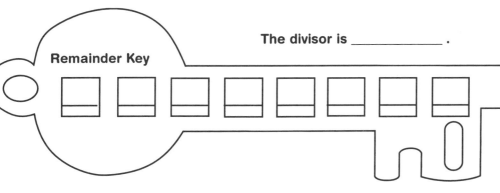

Remainder Key

The divisor is _____ .

Hundred Chart

1	2	3	4	5	6	7	8	9	10
11	12	13	14	15	16	17	18	19	20
21	22	23	24	25	26	27	28	29	30
31	32	33	34	35	36	37	38	39	40
41	42	43	44	45	46	47	48	49	50
51	52	53	54	55	56	57	58	59	60
61	62	63	64	65	66	67	68	69	70
71	72	73	74	75	76	77	78	79	80
81	82	83	84	85	86	87	88	89	90
91	92	93	94	95	96	97	98	99	100

Patterns we found while we were doing this activity:

Nombre:

Restos recurrentes

Hoja de registro

Cómo reunir y organizar los datos

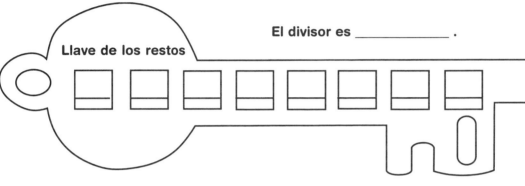

El divisor es _____ .

Llave de los restos

Cuadrícula de centena

1	2	3	4	5	6	7	8	9	10
11	12	13	14	15	16	17	18	19	20
21	22	23	24	25	26	27	28	29	30
31	32	33	34	35	36	37	38	39	40
41	42	43	44	45	46	47	48	49	50
51	52	53	54	55	56	57	58	59	60
61	62	63	64	65	66	67	68	69	70
71	72	73	74	75	76	77	78	79	80
81	82	83	84	85	86	87	88	89	90
91	92	93	94	95	96	97	98	99	100

Patrones que encontramos al realizar esta actividad:

Remainder Rules

Math Concepts

- whole numbers
- multiplication
- division
- subtraction
- addition

Materials

- Math Explorer™
- **Remainder Rules** recording sheets
- cubes, sticks, etc.
- pencils

Overview

Students will use calculators, whole number division, multiplication, addition, and subtraction to generate mathematical expressions that describe the relationships between dividends, divisors, quotients, and remainders.

Introduction

> The **Random Remainders** (page 6) and **Recurring Remainders** (page 10) activities should be completed before beginning this activity.

1. Have students revisit their conclusions from the **Random Remainders** and **Recurring Remainders** activities.

2. Ask students to imagine the following: Suppose you own a flower stand. Each day you order carnations from the dealer and separate them into equal-sized bunches to sell at your stand. You like to take some carnations home each day, so you always place your order and make your bunches so that there are five carnations left for you.

3. Ask students: How many carnations can you order and what size bunches do you make to take five carnations home?

4. Have students work in groups using calculators, cubes, and sticks to explore solutions to this problem. They may also use the sheet of flowers provided on page 17. Have them record their solutions in the chart on the recording sheet.

5. Have students use the data that they collect to describe the relationship between the dividends and divisors that result in the remainder 5. On the recording sheet, have them write mathematical expressions using number and operation symbols that represent their descriptions.

6. Have students repeat the activity using the other charts on the recording sheet. Have them look for similarities in their descriptions.

© 1995 Texas Instruments Incorporated. ™ Trademark of Texas Instruments Incorporated.

Reglas del resto

Conceptos matemáticos

- números enteros
- división
- suma
- multiplicación
- resta

Materiales

- Math Explorer™
- hojas de registro de **Reglas del resto**
- cubos, palos, etc.
- lápices

Resumen

Los alumnos usarán calculadoras, división, multiplicación, suma y resta de números enteros para generar expresiones matemáticas que describan las relaciones entre dividendos, divisores, cocientes y restos.

Introducción

> Antes de comenzar con esta actividad, se deben completar los ejercicios de **Restos al azar** (página 6) y **Restos recurrentes** (página 10).

1. Haga que los alumnos revisen sus conclusiones a partir de las actividades de **Restos al azar** y **Restos recurrentes**.

2. Pida a los alumnos que imaginen la situación siguiente: supongan que son propietarios de un puesto de venta de flores. Todos los días le piden claveles a su distribuidor y los separan en ramos de igual tamaño para venderlos en su puesto. Cada día ustedes se llevan algunos claveles a su casa, de manera que siempre hacen su pedido y preparan los ramos para que sobren cinco claveles.

3. Pregunta para los alumnos: ¿cuántos claveles pueden pedir y de qué tamano son los ramos para que puedan llevarse cinco claveles a su casa?

4. Haga que los alumnos trabajen en grupos usando calculadoras, cubos y palos para probar distintas soluciones a este problema. También pueden usar la hoja de flores que viene en la página 17. Pídales que anoten sus soluciones en el cuadro de la hoja de registro.

5. Haga que los alumnos usen los datos reunidos para describir la relación entre los dividendos y los divisores que dan como resultado el resto cinco. Luego, pídales que usen símbolos numéricos y de operación para escribir expresiones matemáticas que representen sus descripciones en la hoja de registro.

6. Haga que los alumnos repitan la actividad usando los otros cuadros de la hoja de registro. Pídales que busquen similitudes en sus descripciones.

Remainder Rules (continued)

Collecting and Organizing Data

While students are generating and recording their data, ask questions such as:

- What operation represents the action of making bunches of equal size?

- What do the five carnations left over represent?

- What size bunches might you make so that you would have five carnations left?

- How would the size of the bunches you make be used in your description of the action of grouping the flowers in bunches?

- Suppose you made bunches of eight carnations. How many could you order to have zero left?

  Example: 1×8, 2×8, 3×8, etc.

- How would the number of flowers that you order and the size bunches that you make be related to each other if you had none left? How can you express this relationship in words?

  Note: The total number of carnations would have to be a multiple of the number in each bunch.

- How could you express this with symbols?

  Example: If t = total number of carnations and c = carnations in each bunch, for none to be left over, t = some whole number of bunches $\times c$.

- How would the dividend and divisor be related if you wanted to have 1 left?

- How could you express this relationship in symbols?

  Example: t = [some whole number $\times c$] + 1.

How are you using the calculator to help you investigate solutions to this problem?

Which operations are you using on the calculator? Why did you choose those?

How do you record what you are doing with the calculator?

Reglas del resto (continuación)

Cómo reunir y organizar los datos

Mientras los alumnos generan y registran sus datos, haga las preguntas siguientes:

- ¿Qué operación representa la acción de preparar los ramos de igual tamaño?

- ¿Qué representan los cinco claveles que sobran?

- ¿De qué tamaño podrían ser los ramos para que sobraran cinco claveles?

- ¿Cómo se usaría el tamaño de los ramos preparados en su descripción de la acción de agrupar las flores en ramos?

- Supongan que preparan ramos de ocho claveles. ¿Cuántas flores podrían pedir para que no sobrara ninguna?

    Ejemplo: 1×8, 2×8, 3×8, etc.

- ¿Cómo se relacionarían el número de flores que piden y el tamaño de los ramos preparados, si no sobrara ninguna flor? ¿Cómo pueden expresar esta relación en palabras?

    Nota: el número total de claveles debería ser múltiplo de la cantidad de claveles en cada ramo.

- ¿Cómo expresarían esto con símbolos?

    Ejemplo: si t = número total de claveles y c = claveles en cada ramo, para que no sobre ninguno, t = un número entero de ramos x c.

- ¿Cómo se relacionarían el dividendo y el divisor si quisieran que sobrara 1?

- ¿Cómo expresarían esta relación en símbolos?

    Ejemplo: t = [un número entero x c] + 1

- ¿Cómo están usando la calculadora para investigar las soluciones de este problema?

- ¿Qué operaciones están usando en la calculadora? ¿Por qué escogieron esas operaciones?

- ¿Cómo registran lo que están haciendo con la calculadora?

Remainder Rules (continued)

Analyzing Data and Drawing Conclusions

After students have generated several solutions to the problem and looked for a general pattern, have them analyze their data. Ask questions such as:

- What solutions did you find?

- How are your solutions the same or different from those of other groups?

- What strategies did you use in looking for solutions?

- What patterns did you find in your solutions?

- How did you use words, numbers, and operation symbols to describe those patterns?

- How are your descriptions alike or different from those written by others?

- What general statements can you make about a division equation that has a remainder of five?

How did you use the calculator to help you in this activity?

Using the patterns you observed in your data, make a conjecture about another possible solution. Use the calculator to test your conjecture.

Continuing the Investigation

Have students:

- Repeat the activity with this change: Suppose you want to keep eight carnations each day. How are your solutions different from having five left? How are they the same?

- Make a generalization about the relationship between dividends and divisors and any size remainder.

Reglas del resto (continuación)

Cómo analizar los datos y sacar conclusiones

Después de que los alumnos generen varias soluciones al problema y busquen un patrón general, hágalos analizar sus datos. Haga las preguntas siguientes:

- ¿Qué soluciones encontraron?

- ¿En qué se parecen o diferencian sus soluciones y las de los otros grupos?

- ¿Qué estrategias usaron para buscar soluciones?

- ¿Qué patrones encontraron en sus soluciones?

- ¿Cómo usaron las palabras, los números y los símbolos de operación para describir esos patrones?

- ¿En qué se parecen o diferencian sus descripciones y las de los demás?

- ¿Qué afirmaciones generales podrían formular acerca de una ecuación de división que tiene un resto de cinco?

¿Cómo usaron la calculadora en esta actividad?

Con los patrones observados en sus datos, formulen una conjetura sobre otra solución posible. Usen la calculadora para comprobar su conjetura.

Cómo continuar la investigación

Los alumnos deben:

- Repetir la actividad con este cambio: supongan que desean separar ocho claveles cada día. ¿En qué difieren sus soluciones de cuando sobraban cinco? ¿En qué se parecen?

- Hacer una generalización sobre la relación entre los dividendos y los divisores y un resto de cualquier magnitud.

Name:

Remainder Rules
Recording Sheet

Collecting and Organizing Data

Note: t = total number of carnations, and c = number of carnations in each bunch.

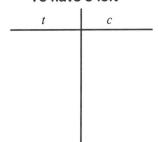

To have 5 left

t	c

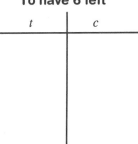

To have 6 left

t	c

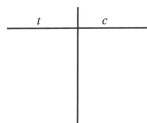

To have 10 left

t	c

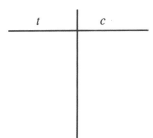

To have n left

t	c

Analyzing Data and Drawing Conclusions

From the data we gathered, we think that the relationship between the dividend (t), the divisor (c), and the remainder is:

Nombre:

Reglas del resto
Hoja de registro

Cómo reunir y organizar los datos

Nota: t = número total de claveles y c = número de claveles en cada ramo.

Para que sobren 5

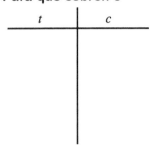

Para que sobren 6

Para que sobren 10

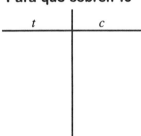

Para que sobren n

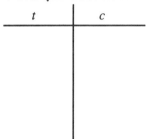

Cómo analizar los datos y sacar conclusiones

A partir de los datos reunidos, pensamos que la relación entre el dividendo (t), el divisor (c) y el resto es la siguiente:

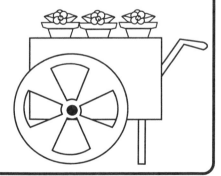

Remainder Rules

Flowers

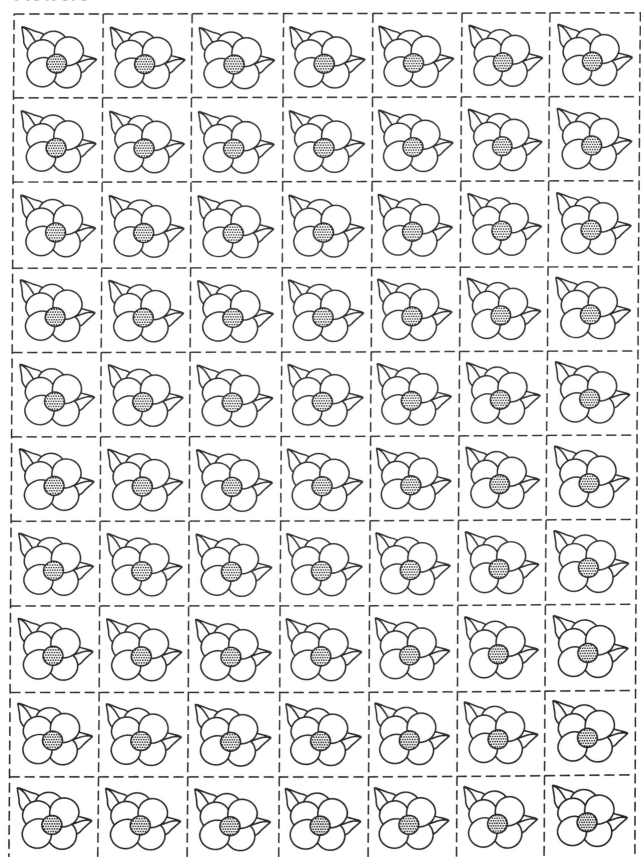

Uncovering Mathematics with Manipulatives and Calculators Levels 2 & 3

Reglas del resto

Flores

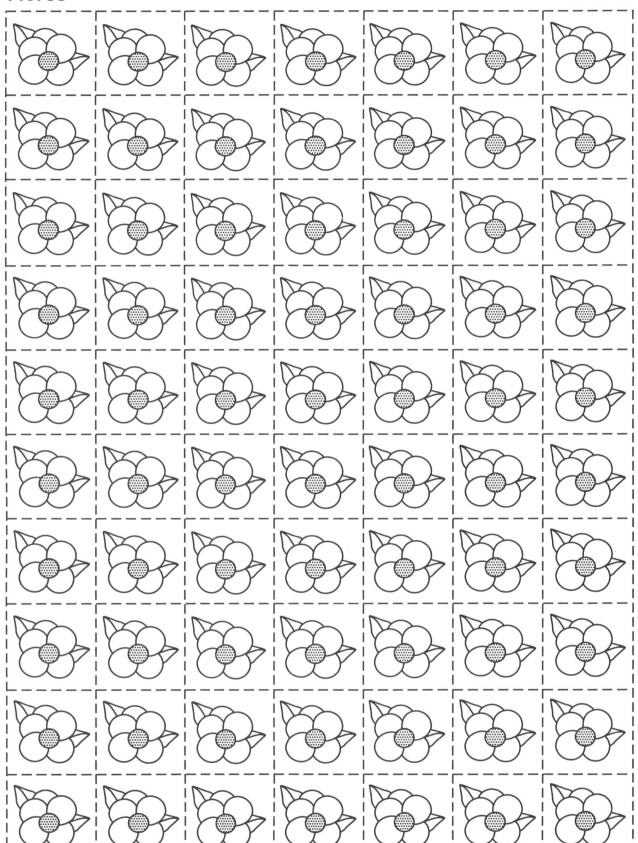

Names for One-Half

Math Concepts

- integers
- fractions
- decimals
- addition
- subtraction
- multiplication
- division

Materials

- Math Explorer™
- **Names for One-Half** recording sheets
- pencils

Overview

Students will use the calculator and their understanding of integers, fractions, decimals, and operations to find mathematical expressions that equal 1/2.

Introduction

1. Discuss situations where being able to express a quantity in several different ways is useful.

   Examples: Two students share a six-pack of soft drinks. Each drinks 3/6 of the six-pack. A child eats 1/4 of a cake on Monday and 1/4 of the same cake on Tuesday. The child has eaten 1/4 + 1/4 of the cake in all.

2. Ask students: How many different names can you find for one-half? (See examples on page 19.)

3. Have students work in pairs. Ask them to use a calculator to find and record as many names for one-half as they can.

Collecting and Organizing Data

While students are exploring with their calculators, ask questions such as:

- What operations are you using?

- What operations have you not used? Why? How could you use those operations?

- How could you make an expression with more than one operation?

- What fractions do you think you could use? How would you use them?

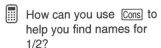 How can you use [Cons] to help you find names for 1/2?

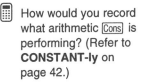 How would you record what arithmetic [Cons] is performing? (Refer to **CONSTANT-ly** on page 42.)

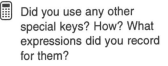 Did you use any other special keys? How? What expressions did you record for them?

Equivalentes para un medio

Conceptos matemáticos

- enteros
- fracciones
- decimales
- suma
- resta
- multiplicación
- división

Materiales

- Math Explorer™
- hojas de registro de **Equivalentes para un medio**
- lápices

Resumen

Los alumnos usarán la calculadora y sus conocimientos de enteros, fracciones, decimales y operaciones para encontrar expresiones matemáticas que sean iguales a 1/2.

Introducción

1. Analice situaciones en las que sea útil poder expresar una cantidad en varias formas diferentes.

   Ejemplos: dos alumnos comparten un paquete de seis bebidas gaseosas. Cada uno toma 3/6 del paquete. Un niño se come 1/4 de una torta el lunes y 1/4 el martes. El niño se comió en total 1/4 + 1/4 de la torta.

2. Pregunta para los alumnos: ¿cuántos nombres distintos pueden encontrar para un medio? (ver ejemplos en la página 19).

3. Con una calculadora, haga que los alumnos trabajen en parejas para encontrar y anotar la mayor cantidad de nombres posibles para un medio.

Cómo reunir y organizar los datos

Mientra los alumnos prueban con sus calculadoras, haga las preguntas siguientes:

- ¿Qué operaciones están usando?

- ¿Qué operaciones no han usado? ¿Por qué? ¿Cómo podrían usar estas operaciones?

- ¿Cómo podrían elaborar una expresión con más de una operación?

- ¿Qué fracciones piensan que podrían usar? ¿Cómo las usarían?

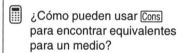 ¿Cómo pueden usar [Cons] para encontrar equivalentes para un medio?

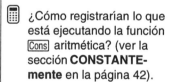 ¿Cómo registrarían lo que está ejecutando la función [Cons] aritmética? (ver la sección **CONSTANTE-mente** en la página 42).

 ¿Usaron alguna otra tecla especial? ¿Cómo? ¿Qué expresiones registraron para esas teclas?

Names for One-Half (continued)

Analyzing Data and Drawing Conclusions

After students have recorded their names for one-half, have them analyze the expressions as a whole group. Ask questions such as:

- How are some of your expressions alike? How are they different?

- If you had to group your expressions, what categories would you use? Why?

- Select one of your categories and see if you can write more expressions that fit that category.

- Choose one of your expressions and describe a real-life situation in which it might be used.

 How did you use the calculator to help you organize your search?

Continuing the Investigation

Have students come up with a class set of categories. Post each category on a wall or chalkboard and have students continue to add expressions.

Examples:

Names for One-half
Names that are
equivalent fractions:

$$\frac{18}{36} \qquad \frac{124}{248}$$

Names for One-half
Names that are
decimals:

0.5

2 – 1.5

0.500

Names for One-half
Names that use
addition:

0.25 + 0.25

$$\frac{1}{3} + \frac{1}{6}$$

Equivalentes para un medio (continuación)

Cómo analizar los datos y sacar conclusiones

Después de que los alumnos registren los nombres para un medio, pídales que en grupo analicen las expresiones. Haga las preguntas siguientes:

- ¿En qué se parecen algunas de sus expresiones? ¿En qué difieren?

- Si tuvieran que agrupar las expresiones, ¿qué categorías usarían? ¿Por qué?

- Seleccionen una de las categorías y vean si pueden escribir más expresiones que pertenezcan en esa categoría.

- Escojan una de las expresiones y describan una situación real en la que se podría usar.

 ¿Cómo usaron la calculadora para organizar la búsqueda?

Cómo continuar la investigación

Haga que los alumnos elaboren un conjunto general de categorías. Distribuya las categorías en las paredes de la sala y pida a los alumnos que continúen agregando expresiones.

Ejemplos:

Equivalentes para un medio

Nombres que usan fracciones equivalentes:

$$\frac{18}{36} \qquad \frac{124}{248}$$

Equivalentes para un medio

Nombres que usan decimales:

$$0.5$$

$$2 - 1.5$$

$$0.500$$

Equivalentes para un medio

Nombres que usan la suma:

$$0.25 + 0.25$$

$$\frac{1}{3} + \frac{1}{6}$$

Name:

Names for One-Half
Recording Sheet

Collecting and Organizing Data

One-half = _____

One-half = _____

One-half = _____

One-half = _____

One-half = _____

One-half = _____

One-half = _____

One-half = _____

One-half = _____

One-half = _____

Analyzing Data and Drawing Conclusions

- Group your expressions into two or more categories. Explain your categories.

- Choose three of your expressions and describe a real-life situation
  in which each one might be used.

Questions we thought of while we were doing this activity:

Nombre:

Equivalentes para un medio
Hoja de registro

Cómo reunir y organizar los datos

Un medio = _____

Un medio = _____

Un medio = _____

Un medio = _____

Un medio = _____

Un medio = _____

Un medio = _____

Un medio = _____

Un medio = _____

Un medio = _____

Cómo analizar los datos y sacar conclusiones

- Agrupen sus expresiones en dos o más categorías. Expliquen sus categorías.

- Escojan tres expresiones y describan una situación real en la que se podrían usar.

Preguntas que surgieron mientras realizábamos esta actividad:

Patterns in Counting with Decimals

MATH
EXPLORER

Math Concepts

- patterns
- decimals
- comparing decimals
- ordering decimals
- place value
- addition

Materials

- Math Explorer™
- **Patterns in Counting with Decimals** recording sheets
- pencils

Overview

Students will use the calculator to connect concrete and symbolic representations of decimal quantities and to recognize patterns in the number symbols.

Introduction

1. Show a 10×10 grid on the overhead (see recording sheet). Ask students: How many squares are there across? How many down? How many in all?

2. Now ask students: If the entire grid is worth one dollar, how much is each square worth? Help students connect their understanding of one cent to 1/100 of a dollar.

3. Have students work in pairs. Give each student a recording sheet. Have the first partner fill in the first blank after the single, small square with a fraction (1/100) and the second blank with a decimal symbol (.01) for one one-hundredth.

4. Have the first partner use the calculator to count by hundredths in decimal form by entering ⊞ **.01** ⊟⊟⊟ . . . and then label the squares on the 10×10 grid with decimals. Have the second partner count by hundredths in fraction form by entering ⊞ **1** ⊘ **100** ⊟⊟⊟ . . . and then label the squares in another grid with fractions.

5. Challenge the students to find as many patterns as they can in each of the labeled grids and record the patterns on their recording sheets.

6. Also challenge the students to record any connections they see between the two labeled grids.

Patrones en la cuenta de decimales

Conceptos matemáticos

- patrones
- decimales
- comparación de decimales
- ordenación de decimales
- valor de posición
- suma

Materiales

- Math Explorer™
- hojas de registro de **Patrones en la cuenta de decimales**
- lápices

Resumen

Los estudiantes utilizarán la calculadora para relacionar representaciones concretas y simbólicas de cantidades decimales y para reconocer patrones en los símbolos numéricos.

Introducción

1. Muestre una cuadrícula de 10×10 en el proyector de transparencias (ver la hoja de registro). Preguntas para los alumnos: ¿cuántos cuadrados hay hacia el lado? ¿Cuántos hay hacia abajo? ¿Cuántos hay en total?

2. En seguida, pregúnteles lo siguiente: si la cuadrícula entera vale un dólar, ¿cuánto cuesta cada cuadrado? Ayúdeles a relacionar su conocimiento de un centavo con 1/100 de un dólar.

3. Haga que los alumnos trabajen en parejas. Déle a cada uno una hoja de registro. Que el primer alumno llene el primer espacio en blanco después del cuadrado pequeño arriba a la izquierda con una fracción (1/100) y el segundo con un símbolo decimal (.01) para un centésimo.

4. Que el primer alumno use la calculadora para contar por centésimos en forma decimal, ingresando ⊞ **.01** ⊟⊟⊟. . . y luego que llene los cuadrados de la cuadrícula de 10×10 con decimales. Que el segundo alumno cuente por centésimos en forma de fracción, ingresando ⊞ **1** ⊘ **100** ⊟⊟⊟ . . . y luego que llene los cuadrados en otra cuadrícula con fracciones.

5. Plantéeles el desafío de encontrar la mayor cantidad de patrones posibles en cada una de las cuadrículas y de registrar los patrones en sus hojas de registro.

6. Además, pídales que registren cualquier conexión que vean entre las dos cuadrículas.

Patterns in Counting
with Decimals (continued)

Collecting and Organizing Data

While students explore with the 10×10 grid and their calculators, ask questions such as:

- What patterns do you see in the decimal representations?

- What patterns do you see in the fraction representations?

- What connections do you see between the fraction and decimal representations?

▦ What do the numbers that show on the calculator after you press ⊟ tell you?

▦ What did you enter to prepare the calculator to count by hundredths? Why?

▦ If you press the F◌D key, a fraction symbol will change to a decimal symbol and vice versa. Try this and see how it connects to your two grids.

Analyzing Data and Drawing Conclusions

After students have looked for patterns and connections, have them work as a whole group to analyze their observations. Ask questions such as:

- What patterns did you notice in the decimal symbols? In the fraction symbols?

- How are the spaces on the grid and the symbols on the calculator connected?

- How are the two kinds of symbols connected?

- With what number does each grid end? Does that make sense? Why or why not?

▦ How did you set up the calculator to count by hundredths in decimal form? In fraction form? Explain your choices.

▦ How could you use the calculator to count ten squares at a time?

▦ What does using the F◌D represent?

Continuing the Investigation

Have students:

- Investigate the question: How would you set up the calculator to count by tenths? How could counting by tenths be represented on the 10×10 grid if the whole grid represents 1?

- Investigate the question: One percent (1%) is another name for 1/100 or .01. Label another 10×10 grid with percent symbols. What connections do you see between the three grids?

Patrones en la cuenta
de decimales (continuación)

Cómo reunir y organizar los datos

Mientras los alumnos prueban con la cuadrícula de 10×10 y la calculadora, haga las preguntas siguientes:

- ¿Qué patrones ven en las representaciones con decimales?

- ¿Qué patrones ven en las representaciones con fracción

- ¿Qué relaciones ven entre las representaciones con decimales y con fracción?

¿Qué les indican los números que aparecen en la calculadora después de presionar =?

¿Qué ingresaron para preparar la calculadora para contar por centésimos? ¿Por qué?

Si presionan la tecla F⊃D, un símbolo de fracción cambia a símbolo de decimal y viceversa. Prueben esta función y vean cómo se relaciona con sus dos cuadrículas.

Cómo analizar los datos y sacar conclusiones

Después de que los alumnos busquen patrones y relaciones, pídales que trabajen como un solo grupo para analizar sus observaciones. Haga las preguntas siguientes:

- ¿Qué patrones observaron en los símbolos de decimal? ¿Y en los símbolos de fracción?

- ¿Cómo se relacionan los espacios en la cuadrícula y los símbolos en la calculadora?

- ¿Cómo se relacionan los dos tipos de símbolos?

- ¿Con qué número termina cada cuadrícula? ¿Tiene eso algún significado? ¿Por qué?

¿Cómo configuraron la calculadora para contar por centésimos en forma decimal? ¿Y en forma de fracción? Expliquen por qué.

¿Cómo podrían usar la calculadora para contar diez cuadrados a la vez?

¿Qué representa el uso de F⊃D?

Cómo continuar la investigación

Los alumnos deben:

- Investigar las preguntas siguientes: ¿cómo se puede configurar la calculadora para que cuente por décimos? ¿Cómo se podría representar la cuenta por décimos en la cuadrícula de 10×10, si toda la cuadrícula representa 1?

- Investigar la pregunta siguiente: uno por ciento (1%) es otro nombre para 1/100 ó .01. Llene otra cuadrícula de 10 x 10 con símbolos porcentuales. ¿Qué relaciones ven entre las tres cuadrículas?

Patterns in Counting with Decimals

Recording Sheet

Collecting and Organizing Data

10 × 10 Grid

☐ = ————
  Fraction

or   ————
     Decimal

Analyzing Data and Drawing Conclusions

- The patterns we saw in the decimal symbols:

- The patterns we saw in the fraction symbols:

- The connections we saw between the two kinds of symbols:

Nombre:

Patrones en la cuenta de decimales
Hoja de registro

Cómo reunir y organizar los datos

Cuadrícula de 10 x 10

$\boxed{}$ = $\dfrac{}{\text{Fracción}}$

O $\dfrac{}{\text{Decimal}}$

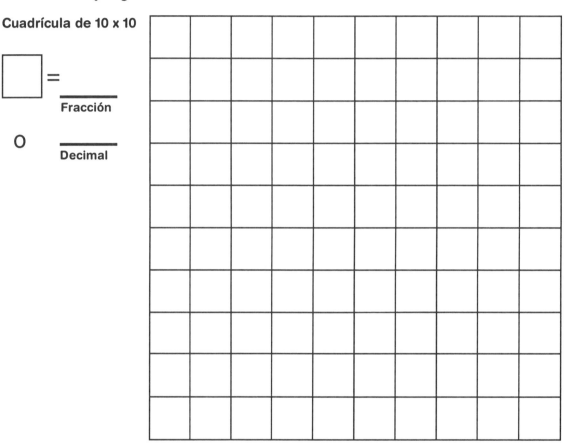

Cómo analizar los datos y sacar conclusiones

• Los patrones que observamos en los símbolos decimales son los siguientes:

• Los patrones que observamos en los símbolos de fracción son los siguientes:

• Las relaciones que observamos entre los dos tipos de símbolos son las siguientes:

Names for 100

Math Concepts

- addition
- subtraction
- multiplication
- division
- decimals
- fractions
- integers

Materials

- Math Explorer™
- **Names for 100** recording sheets
- pencils

Overview

Students will use the calculator and their understanding of integers, fractions, decimals, and operations to find mathematical expressions that equal 100.

Introduction

1. Discuss situations in which being able to express a quantity in several different ways is useful.

    Examples: The number six could be thought of as one more than five (5 + 1) because if each of the five people in your family drinks one canned drink from a six-pack, there is one left. Six could also be thought of as 2 × 3 because six slices of bread are needed to make three sandwiches for lunch.

2. Ask students: How many different names do you think you can find for 100? (See examples on page 25.)

3. Have students work in pairs. Ask them to use a calculator to find and record as many different names for 100 as they can.

Collecting and Organizing Data

While students are exploring with their calculators, ask questions such as:

- What operations are you using?

- What operations have you not used? Why? How could you use those operations?

- How could you make an expression with more than one operation?

- What fractions do you think you could use? How would you use them?

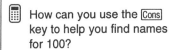 How can you use the [Cons] key to help you find names for 100?

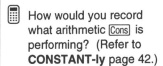 How would you record what arithmetic [Cons] is performing? (Refer to **CONSTANT-ly** page 42.)

Did you use any other special keys? How? What expressions did you record for them?

Equivalentes para la centena

Conceptos matemáticos

- suma
- resta
- multiplicación
- división

- decimales
- fracciones
- enteros

Materiales

- Math Explorer™
- hojas de registro de **Equivalentes para la centena**
- lápices

Resumen

Los alumnos utilizarán la calculadora y sus conocimientos de los enteros, fracciones, decimales y operaciones para encontrar la mayor cantidad posible de expresiones matemáticas equvalentes a 100.

Introducción

1. Examine situaciones en las que sea útil poder expresar una cantidad en varias formas distintas.

   Ejemplos: el número seis se podría considerar como uno más que cinco (5+1) porque si cada una de las cinco personas en su familia toma una bebida de un paquete de seis, sólo sobra una. Seis también se puede considerar como 2x3, porque se necesitan seis rebanadas de pan para hacer tres emparedados para el almuerzo.

2. Pregunta para los alumnos: ¿cuántos nombres creen que pueden encontrar para 100? (ver los ejemplos de la página 25).

3. Haga que los alumnos trabajen en parejas, usando una calculadora, para encontrar y registrar la mayor cantidad posible de nombres diferentes para 100.

Cómo reunir y organizar los datos

Mientras los alumnos prueban con sus calculadoras, haga las preguntas siguientes:

- ¿Qué operaciones están usando?

- ¿Qué operaciones no han usado? ¿Por qué? ¿Cómo podrían usar esas operaciones?

- ¿Cómo podrían hacer una expresión con más de una operación?

- ¿Qué fracciones piensan que podrían usar? ¿Cómo las usarían?

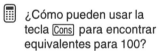 ¿Cómo pueden usar la tecla Cons para encontrar equivalentes para 100?

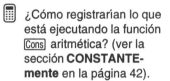 ¿Cómo registrarían lo que está ejecutando la función Cons aritmética? (ver la sección **CONSTANTE-mente** en la página 42).

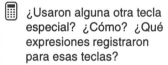 ¿Usaron alguna otra tecla especial? ¿Cómo? ¿Qué expresiones registraron para esas teclas?

Names for 100 (continued)

Analyzing Data and Drawing Conclusions

After students have recorded their names for 100, have them analyze the expressions as a whole group. Ask questions such as:

- How are some of your expressions alike? How are they different?

- If you had to group your expressions, what categories would you use? Why?

- Select one of your categories and see if you can write more expressions that fit that category.

- Choose one of your expressions and describe a real-life situation in which it might be used.

 How did you use the calculator to help you organize your search?

Continuing the Investigation

Have students come up with a class set of categories. Post each category on a chalkboard or wall and have students continue to add expressions.

Examples:

Names for 100
Names that use division:

$500 \div 5$

$50 \div 1/2$

Names for 100
Names that use addition and multiplication:

$50 + 4 (12\ 1/2)$

$50 + 5 (10)$

Names for 100
Names that use 1/2:

$50 \div 1/2$

$49\ 1/2 + 50\ 1/2$

12.5×8

Equivalentes para la centena (continuación)

Cómo analizar los datos y sacar conclusiones

Después de que los alumnos registren sus equivalentes para 100, pídales que analicen las expresiones como un solo grupo. Haga las preguntas siguientes:

- ¿En qué se parecen algunas de sus expresiones? ¿En qué difieren?

- Si tuvieran que agrupar sus expresiones, ¿qué categorías usarían? ¿Por qué?

- Seleccionen una de sus categorías y vean si pueden escribir más expresiones que pertenezcan en ella.

- Escojan una de sus expresiones y describan una situación real en la que se podría usar.

🖩 ¿Cómo usaron la calculadora para organizar su búsqueda?

Cómo continuar la investigación

Que los alumnos elaboren un conjunto general de categorías. Distribuya cada categoría en las paredes de la sala y pida a los alumnos que sigan agregando expresiones.

Ejemplos:

Equivalentes para 100
Nombres que usan
la división:

$500 \div 5$

$50 \div 1/2$

Equivalentes para 100
Nombres que usan la suma
y la multiplicación:

$50 + 4 (12\ 1/2)$

$50 + 5 (10)$

Equivalentes para 100
Nombres que usan 1/2:

$50 \div 1/2$

$49\ 1/2\ +\ 50\ 1/2$

12.5×8

Name:

Names for 100
Recording Sheet

Collecting and Organizing Data

100 = _____

100 = _____

100 =_____

100 = _____

100 = _____

100 = _____

100 = _____

100 =_____

100 = _____

100 = _____

Analyzing Data and Drawing Conclusions

• Group your expressions into two or more categories. Explain your categories.

• Choose three of your expressions and describe a real-life situation in which each one might be used.

Questions we thought of while we were doing this activity:

Nombre:

Equivalentes para la centena
Hoja de registro

Cómo reunir y organizar los datos

100 = _____

100 = _____

100 = _____

100 = _____

100 = _____

100 = _____

100 = _____

100 = _____

100 = _____

100 = _____

100 = _____

Cómo analizar los datos y sacar conclusiones

- Agrupe sus expresiones en dos o más categorías. Explique sus categorías.

- Escoja tres de sus expresiones y describa una situación real en la que se podría usar cada una.

Preguntas que surgieron mientras realizábamos esta actividad:

Activities:

1   Area Patterns

2   Perimeter Patterns

3   The Mysterious Constant

4   CONSTANT-ly

5   "Power"ful Patterns

6   Patterns in Division

7   Picturing Percents

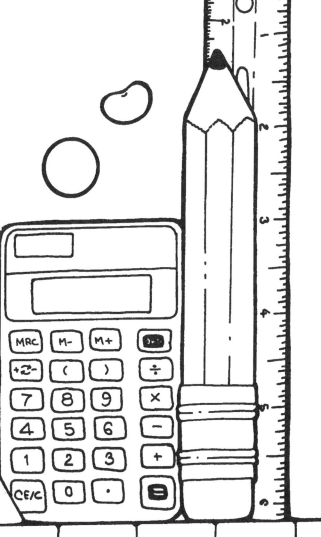

Area Patterns

Math Concepts

- whole numbers
- comparing numbers
- estimation
- measuring area
- addition
- multiplication
- functions
- similar shapes

Materials

- TI-108, Math Mate™, Math Explorer™
- Pattern Blocks
- **Area Patterns** recording sheets
- pencils

Overview

Students will investigate patterns in ordered pairs generated by constructing a sequence of similar shapes. They will then use the patterns and the calculator to predict the number of blocks it will take to build a specific shape in the sequence.

Introduction

1. Have students use the green triangles from Pattern Blocks (or the paper triangle provided on page 32) to make the following pattern.

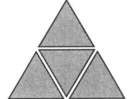

2. Ask students to predict how many green triangles it will take to make the next larger triangle of the same shape. Continue the pattern.

3. In the table on the recording sheet, have students draw each triangle and then record the number of blocks it took to make it (its area in green triangles).

4. Have students investigate the patterns in their tables, use the calculator to predict the area of the 95th triangle, and write their predictions on the recording sheet.

5. Have students choose a different Pattern Block (such as the blue rhombus) and perform the same investigation.

6. Ask students to compare the patterns generated by the different shapes and write about their discoveries.

Patrones de superficie

Conceptos matemáticos

- números enteros
- comparación numérica
- estimación
- medición de superficie
- figuras geométricas similares
- suma
- multiplicación
- funciones

Materiales

- TI-108, Math Mate™, Math Explorer™
- Figuras Patrón
- hojas de registro de **Patrones de superficie**
- lápices

Resumen

Los estudiantes investigarán patrones en pares ordenados generados al construir una secuencia de figuras geométricas similares. Después, utilizarán los patrones y la calculadora para predecir el número de figuras que se necesitarán para construir una figura geométrica específica en la secuencia.

Introducción

1. Haga que los alumnos usen los triángulos verdes a partir de las Figuras Patrón (o el triángulo de papel que viene en la página 32) para construir el patrón siguiente:

 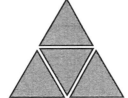 

2. Que los alumnos predigan cuántos triángulos verdes se necesitarán para hacer el tríangulo que sigue en tamaño de la misma figura geométrica. Que continúen el patrón.

3. En la hoja de registro, que los alumnos dibujen cada triángulo y luego registren en el cuadro la cantidad de figuras que se necesitaron para hacerlo (su superficie en triángulos verdes).

4. Que los alumnos investiguen los patrones en sus cuadros, y que usen la calculadora para predecir la superficie del 95º triángulo y escriban sus predicciones en la hoja de registro.

5. Que los alumnos escojan una Figura Patrón diferente (por ejemplo, el rombo azul) y ejecuten la misma investigación.

6. Que los alumnos comparen los patrones generados por las distintas figuras geométricas y escriban sobre sus descubrimientos.

Area Patterns (continued)

Collecting and Organizing Data

While students explore their patterns, ask questions such as:

- What unit of area are you using to measure the area of each shape? Why do you think it is an effective unit of measure?

- What do the numbers in your table(s) represent?

- What patterns do you notice in your table(s)?

- How can you be sure you made the next larger triangle? Do the patterns in your table help you discover when you have skipped a triangle? How?

▢ How are the numbers you see in the calculator display connected to the numbers in your table(s)?

▢ How can you use the calculator to predict the number of blocks it will take to construct the 95th shape in your sequence?

▢ What happens if your numbers get too big?

Analyzing Data and Drawing Conclusions

After students have investigated several sequences with different pattern blocks, have them work as a whole group to analyze the patterns in the ordered pairs in their tables. Ask questions such as:

- How is your table for the green triangle different from your table for the blue rhombus? Why do you think they are different?

- Are any of your tables alike? How can you explain this?

- What difficulties did you have with the red trapezoid? How did you handle these problems? How did the table for the red trapezoid compare with the tables for some of the other shapes you investigated?

- What patterns did you notice in your tables? How did you describe these patterns?

- What discoveries did you make?

▢ How did you use your calculator to help you make predictions?

▢ How did you use your calculator to discover the patterns in the ordered pairs in your table(s)?

© 1995 Texas Instruments Incorporated. ™ Trademark of Texas Instruments Incorporated.

Patrones de superficie (continuación)

Cómo reunir y organizar los datos

Mientras los alumnos prueban sus patrones, haga las preguntas siguientes:

- ¿Qué unidad de superficie están usando para medir la superficie de cada figura geométrica? ¿Por qué piensan que es una unidad de medida eficaz?

- ¿Qué representan los números en sus cuadros?

- ¿Qué patrones observan en sus cuadros?

- ¿Cómo pueden estar seguros de que hicieron el triángulo que seguía en tamaño? ¿Los patrones en su cuadro les ayudan a determinar cuándo omitieron un triángulo? ¿Cómo?

⊞ ¿Cómo se relacionan los números que ven en el visor de la calculadora con los de sus cuadros?

⊞ ¿Cómo pueden usar la calculadora para predecir el número de figuras que se necesitarán para construir la 95ª forma en su secuencia?

⊞ ¿Qué sucede si sus números se hacen demasiado grandes?

Cómo analizar los datos y sacar conclusiones

Después de que los alumnos investiguen varias secuencias con distintas figuras patrón, pídales que analicen como un solo grupo los patrones en los pares ordenados de sus cuadros. Haga las preguntas siguientes:

- ¿En qué difieren su cuadro del triángulo verde y el del rombo azul? ¿Por qué creen que son diferentes?

- ¿Algunos de sus cuadros se parecen? ¿Cómo pueden explicar esta situación?

- ¿Qué dificultades tuvieron con el trapezoide rojo? ¿Cómo manejaron estos problemas? ¿Cómo se compara el cuadro del trapezoide rojo con los cuadros de algunas de las demás figuras geométricas que investigaron?

- ¿Qué patrones observaron en sus cuadros? ¿Cómo describieron esos patrones?

- ¿Qué descubrimientos hicieron?

⊞ ¿Cómo usaron la calculadora para hacer las predicciones?

⊞ ¿Cómo usaron la calculadora para descubrir los patrones en los pares ordenados de sus cuadros?

Area Patterns (continued)

Continuing the Investigation

Have students:

- Choose shapes not included in the Pattern Blocks (such as a rectangle or right triangle on page 32). Use the patterns in the ordered pairs generated by the Pattern Blocks to investigate patterns generated by these other shapes.

- Generate a table of ordered pairs and see if they can find a series of shapes to go with it.

Patrones de superficie (continuación)

Cómo continuar la investigación

Los alumnos deben:

- Escoger figuras geométricas no incluidas en las Figuras Patrón (por ejemplo, un rectángulo o un triángulo restángulo de la página 32). Deben usar los patrones en los pares ordenados generados por las Figuras Patrón para investigar los patrones generados por esas otras figuras geométricas.

- Generar un cuadro de pares ordenados y ver si pueden encontrar una serie de figuras geométricas que lo acompañen.

Name:

Area Patterns
Recording Sheet

Collecting and Organizing Data

Our first four or five similar shapes:

Our data is recorded here:

Shapes	Area (# of blocks)
1 (st)	
2 (nd)	
3 (rd)	
4 (th)	
5 (th)	
6 (th)	
•	
•	
•	

Analyzing Data and Drawing Conclusions

- A pattern we discovered in our table is:

- The 95th shape will take _____ blocks to build. We think this because:

Questions we thought of while we were doing this activity:

Nombre:

Patrones de superficie
Hoja de registro

Cómo reunir y organizar los datos

Hagan un dibujo de las primeras cuatro o cinco figuras geométricas similares en el espacio siguiente:

Registren sus datos:

Figuras geométricas	Superficie (No. de figuras)
1ª	
2ª	
3ª	
4ª	
5ª	
6ª	
•	
•	
•	

Cómo analizar los datos y sacar conclusiones

- Describan un patrón que hayan descubierto en su cuadro.

- Para crear la 95ª figura geométrica, se necesitan _____ figuras. Pensamos así por las razones siguientes:

Preguntas que surgieron mientras realizábamos esta actividad:

Area Patterns

Pattern Blocks

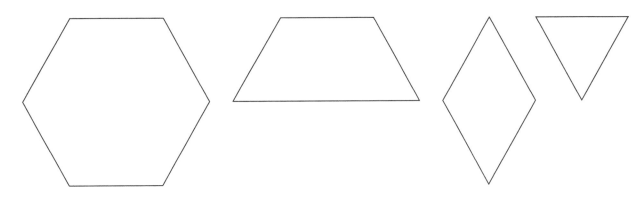

- -

Other Geometric Shapes

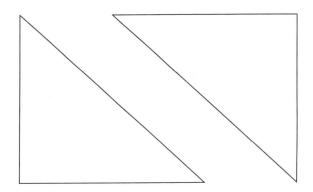
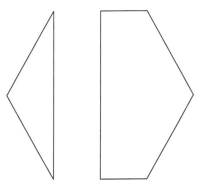

Patrones de superficie

Figuras Patrón

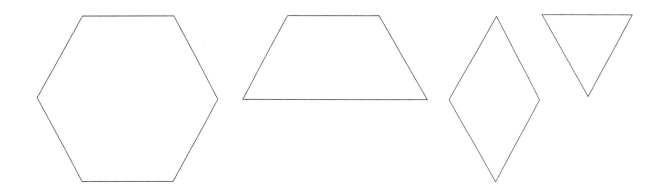

Otras figuras geométricas

Perimeter Patterns

Math Concepts

- whole numbers
- comparing numbers
- estimation
- measuring perimeter
- addition
- multiplication
- functions
- similar shapes

Materials

- TI-108, Math Mate™, Math Explorer™
- Pattern Blocks
- **Perimeter Patterns** recording sheets
- pencils

Overview

Students will investigate patterns in ordered pairs generated by constructing a sequence of similar shapes. They will then use the patterns and the calculator to predict the perimeter of a specific shape in the sequence.

Introduction

> The **Area Patterns** activity on page 28 should be completed before beginning this activity.

1. Have students use green triangles from Pattern Blocks (or the paper triangle provided on page 37) to make the following pattern.

 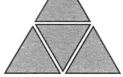

2. Have students figure the perimeter of each triangle (each side of the small triangle counts as one unit) and predict the perimeter of the next triangle. Continue the pattern.

3. Have students draw each triangle and then record its perimeter in the table on the recording sheet.

4. Have students investigate the patterns in their tables, use the calculator to predict the perimeter of the 95th triangle, and write the prediction on the recording sheet.

5. Have students choose a different Pattern Block (such as the blue rhombus) and perform the same investigation.

6. Have students compare the patterns generated by the different shapes and write about their discoveries.

Patrones de perímetro

TI-108
MATH MATE
MATH
EXPLORER

Conceptos matemáticos

- números enteros
- comparación numérica
- estimación
- medición de perímetro
- figuras geométricas similares

- suma
- multiplicación
- funciones

Materiales

- TI-108, Math Mate™, Math Explorer™
- Figuras Patrón
- hojas de registro de **Patrones de perímetro**
- lápices

Resumen

Los alumnos investigarán patrones en pares ordenados generados al construir una secuencia de formas geométricas similares. Después, usarán los patrones y la calculadora para predecir el perímetro de una figura geométrica específica en la secuencia.

Introducción

> Antes de comenzar con esta actividad, se debe completar el ejercicio de **Patrones de superficie** de la página 28.

1. Que los alumnos usen triángulos verdes a partir de las Figuras Patrón (o el triángulo de papel que viene en la página 37) para hacer el patrón siguiente.

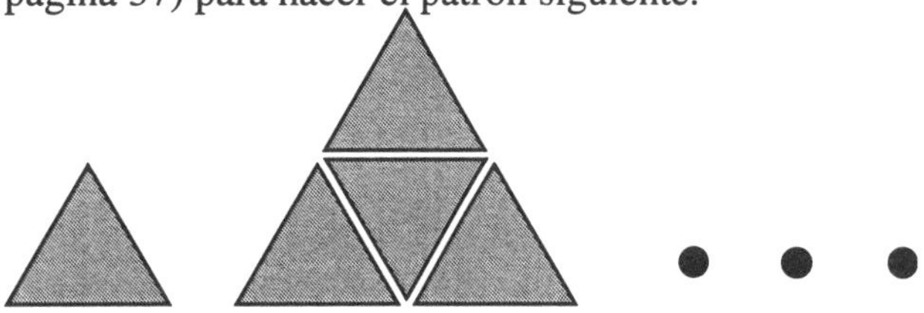

2. Que los alumnos estimen el perímetro de cada triángulo (cada lado del triángulo pequeño cuenta como una unidad) y que predigan el perímetro del triángulo siguiente. Que continúen el patrón.

3. Que los alumnos dibujen cada triángulo y luego registren su perímetro en el cuadro de la hoja de registro.

4. Que los alumnos investiguen los patrones en sus cuadros, usen la calculadora para predecir el perímetro del 95º triángulo y escriban la predicción en la hoja de registro.

5. Que los alumnos escojan una Figura Patrón diferente (por ejemplo, el rombo azul) y ejecuten la misma investigación.

6. Que los alumnos comparen los patrones generados por las diferentes figuras geométricas y escriban sobre sus descubrimientos.

Perimeter Patterns (continued)

Collecting and Organizing Data

While students explore their patterns, ask questions such as:

- What unit are you using to measure the perimeters? Why do you think it is an effective unit of measure to use?

- What do the numbers in your table(s) represent?

- What patterns do you notice in your table(s)?

- How can you be sure you made the next larger triangle? Do the patterns in your tables help you discover when you have skipped a triangle? How?

How are the numbers you see in the calculator display connected to the numbers in your table(s)?

How can you use the calculator to predict the perimeter of the 95th shape in your sequence?

What happens if your numbers get too big?

Analyzing Data and Drawing Conclusions

After students have investigated several sequences with different Pattern Blocks, have them work as a whole group to analyze the patterns in the ordered pairs in their tables. Ask questions such as:

- How is your table for the green triangle different from your table for the orange square or the blue rhombus? Why do you think they are different?

- Are any of your tables alike? How can you explain this?

- What difficulties did you have with the red trapezoid? How did you handle these problems? How did the table for the red trapezoid compare with the tables of some of the other shapes you investigated?

- What patterns did you notice in your tables? How did you describe these patterns?

- What discoveries did you make?

- How did the patterns in this activity compare with the **Area Patterns** (page 28) you discovered?

How did you use your calculator to help you make predictions?

How did you use your calculator to discover the patterns in the ordered pairs in your table(s)?

© 1995 Texas Instruments Incorporated. ™ Trademark of Texas Instruments Incorporated.

Patrones de perímetro (continuación)

Cómo reunir y organizar los datos

Mientras los alumnos prueban sus patrones, haga las preguntas siguientes:

- ¿Qué unidad están usando para medir los perímetros? ¿Por qué piensan que es una unidad de medida eficaz?

- ¿Qué representan los números en sus cuadros?

- ¿Qué patrones observan en sus cuadros?

- ¿Cómo pueden estar seguros de que hicieron el triángulo que seguía en tamaño? ¿Los patrones en su cuadro les ayudan a determinar cuándo han omitido un triángulo? ¿Cómo?

▦ ¿Cómo se relacionan los números que ven en el visor de la calculadora y los de sus cuadros?

▦ ¿Cómo pueden usar la calculadora para predecir el perímetro de la 95ª figura geométrica en su secuencia?

▦ ¿Qué sucede si sus números se hacen demasiado grandes?

Cómo analizar los datos y sacar conclusiones

Después de que los alumnos investiguen varias secuencias con diferentes Figuras Patrón, pídales que analicen como un solo grupo los patrones en los pares ordenados de sus cuadros. Haga las preguntas siguientes:

- ¿En qué difieren su cuadro del triángulo verde y el del cuadrado anaranjado o el rombo azul? ¿Por qué piensan que son diferentes?

- ¿Algunos de los cuadros se parecen? ¿Cómo pueden explicar esta situación?

- ¿Qué dificultades tuvieron con el trapezoide rojo? ¿Cómo manejaron estos problemas? ¿Cómo se comparan el cuadro del trapezoide rojo y los de algunas de las otras figuras geométricas que investigaron?

- ¿Qué patrones observaron en sus cuadros? ¿Cómo describieron esos patrones?

- ¿Qué descubrimientos hicieron?

- ¿Cómo se comparaban los patrones en esta actividad con los **Patrones de superficie** (página 28) que ustedes descubrieron?

▦ ¿Cómo usaron la calculadora para hacer las predicciones?

▦ ¿Cómo usaron la calculadora para descubrir los patrones en los pares ordenados de sus cuadros?

Perimeter Patterns (continued)

Continuing the Investigation

Have students:

- Choose shapes not included in the Pattern Blocks (such as a rectangle or right triangle on page 37). Identify a unit to measure the perimeters and look for patterns.

- Generate a table of ordered pairs and see if they can find a series of shapes to go with it.

Patrones de perímetro (continuación)

Cómo continuar la investigación

Los alumnos deben:

- Escoger figuras geométricas no incluidas en las Figuras Patrón (por ejemplo, un rectángulo o un triángulo rectángulo de la página 37). Deben identificar una unidad para medir los perímetros y buscar patrones.

- Generar un cuadro de pares ordenados y ver si pueden encontrar una serie de figuras geométricas que lo complementen.

Name:

Perimeter Patterns
Recording Sheet

Collecting and Organizing Data

Our first four or five similar shapes:

Our data is recorded here:

Shapes	Perimeter
1 (st)	
2 (nd)	
3 (rd)	
4 (th)	
5 (th)	
6 (th)	
.	
.	
.	

Analyzing Data and Drawing Conclusions

- A pattern we discovered in our table is:

- The 95th shape will have a perimeter of _____. We know this because:

Questions we thought of while we were doing this activity:

Nombre:

Patrones de perímetro
Hoja de registro

Cómo reunir y organizar los datos

Hagan un dibujo de las primeras cuatro o cinco figuras geométricas similares en el espacio siguiente:

Registren sus datos:

Figuras geométricas	Perímetro
1ª	
2ª	
3ª	
4ª	
5ª	
6ª	
•	
•	
•	

Cómo analizar los datos y sacar conclusiones

*   Describan un patrón que hayan descubierto en su cuadro.

*   La 95ª figura geométrica tendrá un perímetro de _____. Sabemos esto por las razones siguientes:

Preguntas que surgieron mientras realizábamos esta actividad:

Pattern Blocks

Pattern Blocks

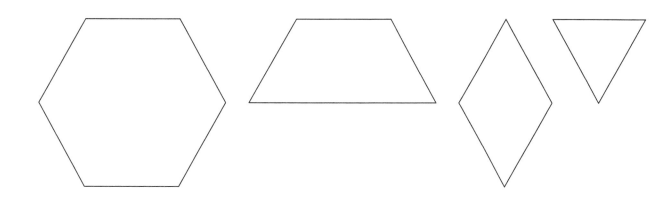

Other Geometric Shapes

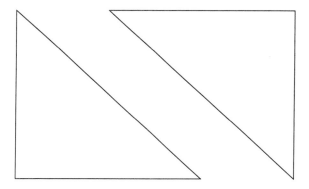

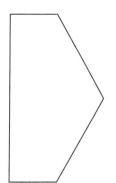

Figuras patrón

Figuras Patrón

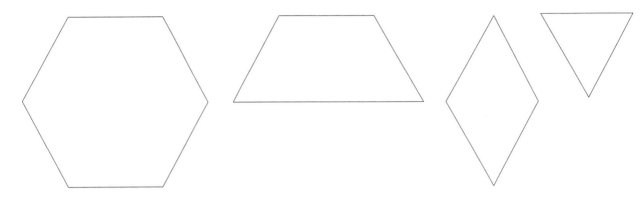

Otras figuras geométricas

The Mysterious Constant

Math Concepts

- whole numbers
- comparing numbers
- estimation
- addition
- multiplication
- functions

Materials

- Math Explorer™
- **The Mysterious Constant** recording sheets
- pencils

Overview

Students will investigate a pattern using the constant function on the calculator, record the ordered pairs in a table, describe the pattern and predict what will come next.

Introduction

1. Have students work in pairs to play the following game. The first partner will secretly write and then enter a constant function and a starting number on the calculator. Then he or she will press Cons once more and hand the calculator to the second partner.

   Example: The first partner might write and then enter
   ⊟ 2 Cons 100 Cons .

2. The second partner will press Cons repeatedly. He or she will record the displayed value in the table on the recording sheet after each press of the Cons key.

Example:

Number of times Cons has been pressed (left side of the display)	Result (right side of the display)
1	98
2	96
3	94
4	92
5	90
6	88
7	86
8	84
9	82
10	80

La misteriosa constante

Conceptos matemáticos

- números enteros
- comparación numérica
- estimación
- suma
- multiplicación
- funciones

Materiales

- Math Explorer™
- hojas de registro de **La misteriosa constante**
- lápices

Resumen

Los alumnos investigarán un patrón usando la función constante de la calculadora, registrarán los pares ordenados en un cuadro, describirán el patrón y pronosticarán lo que sucederá a continuación.

Introducción

1. Haga que los alumnos trabajen en parejas para jugar el siguiente juego. El primer alumno escribirá e ingresará, en secreto, una función constante y un número inicial en la calculadora. Luego, presionará la tecla [Cons] una vez más y le pasará la calculadora al segundo alumno.

   Ejemplo: [−] **2** [Cons] **100** [Cons].

2. El segundo alumno presionará repetidamente la tecla [Cons] y cada vez registrará el valor desplegado en el cuadro de la hoja de registro.

   Ejemplo:

Número de veces que se presiona [Cons] (lado izquierdo del visor)	Resultado (lado derecho del visor)
1	98
2	96
3	94
4	92
5	90
6	88
7	86
8	84
9	82
10	80

The Mysterious Constant (continued)

Introduction (continued)

3. Have each pair of students study the pattern in their table and describe it on the recording sheet. Have the student who built the table guess the mystery constant function and predict what will come next in the pattern. Have both partners in each pair predict the number that would be displayed on the 95th press of Cons and write it on the recording sheet.

4. Have partners switch roles and continue the same investigation with a different mystery constant function.

Collecting and Organizing Data

While students explore their patterns, ask questions such as:

*   Are the numbers in your table getting larger or smaller? By how much?

*   What do the numbers in your table represent?

*   What patterns do you notice in your table?

*   What would happen if the mystery constant function remained the same, but the starting number were different? How would the numbers in your table change?

*   What would happen if you entered a constant function with a negative number? With a fraction? With a decimal?

How could you describe what the calculator does each time you press Cons? What should come next in your table if you are correct?

How can you use the calculator to predict the 95th number in your table?

What happens if your numbers get too big?

Analyzing Data and Drawing Conclusions

After students have investigated several mystery constant functions as a group, have them work as a whole group to analyze the patterns in the ordered pairs in their tables. Ask questions such as:

*   What are some strategies you used to guess the mystery constant function?

*   Are any of your sets of data like those of another group? How can you explain this?

How did you use your calculator to help you make predictions?

© 1995 Texas Instruments Incorporated. ™ Trademark of Texas Instruments Incorporated.

La misteriosa constante (continuación)

Introducción (continuación)

3. Que cada pareja estudie el patrón en su tabla y lo describa en la hoja de registro. Que el alumno que creó el cuadro adivine la función constante misteriosa y haga una predicción de lo que pasará a continuación en el patrón. Que los dos alumnos de cada pareja predigan el número que se desplegaría la 95ª vez que se presione [Cons] y lo escriba en la hoja de registro.

4. Que los alumnos intercambien papeles y continúen la misma investigación con una función constante misteriosa diferente.

Cómo reunir y organizar los datos

Mientras los alumnos prueban sus patrones, haga las preguntas siguientes:

- ¿Los números de su cuadro se hacen más grandes o más pequeños? ¿En qué proporción?

- ¿Qué representan los números de su cuadro?

- ¿Qué patrones observan en su cuadro?

- ¿Que sucedería si la función constante misteriosa siguiera siendo la misma, pero el número inicial fuera diferente? ¿Cómo cambiarían los números de su cuadro?

- ¿Qué sucedería si ingresara una función constante con un número negativo? ¿Con una fracción? ¿Con un decimal?

🔢 ¿Cómo describirían lo que hace la calculadora cada vez que presionan [Cons]? ¿Qué debería venir después en el cuadro si tienen razón?

🔢 ¿Cómo pueden utilizar la calculadora para predecir el 95º número en el cuadro?

🔢 ¿Qué sucede si los números se vuelven demasiado grandes?

Cómo analizar los datos y sacar conclusiones

Después de que los alumnos investiguen varias funciones constantes misteriosas diferentes, hágalos trabajar como un solo grupo para analizar los patrones en los pares ordenados en sus cuadros. Haga las preguntas siguientes:

- ¿Cuáles son algunas de las estrategias que utilizaron para adivinar la función constante misteriosa?

- ¿Algunos de sus conjuntos de datos se parecen a los de otro grupo? ¿Cómo pueden explicar esta situación?

🔢 ¿Cómo utilizaron la calculadora para hacer sus predicciones?

The Mysterious Constant (continued)

Analyzing Data and Drawing Conclusions (continued)

- What patterns did you notice in your tables? How did you describe these patterns?

- Which operations were the most difficult to guess in your mystery constant functions? Why do you think this is true?

- What discoveries did you make?

- How did the patterns in this activity compare with the patterns you discovered in **Perimeter Patterns** (page 33) and **Area Patterns** (page 28)?

How did you use your calculator to discover the patterns in the ordered pairs in your table(s)?

Continuing the Investigation

Have students:

- Choose a pattern from **Perimeter Patterns** (page 33) and **Area Patterns** (page 28) and see if you can enter a constant function to duplicate the pattern in that table of ordered pairs.

- Choose one of your tables of ordered pairs and see if you can think of a real-life situation it might represent.

La misteriosa constante (continuación)

Cómo analizar los datos y sacar conclusiones (continuación)

- ¿Qué patrones observaron en sus cuadros? ¿Cómo describieron esos patrones?

- ¿Qué operaciones fueron más difíciles de adivinar en sus funciones constantes misteriosas? ¿Por qué creen que sucede esto?

- ¿Qué descubrimientos hicieron?

- ¿Cómo se comparaban los patrones de esta actividad con los descubiertos en **Patrones de perímetro** (página 33) y **Patrones de superficie** (página 28)?

 ¿Cómo utilizaron la calculadora para descubrir los patrones en los pares ordenados de sus cuadros?

Cómo continuar la investigación

Los alumnos deben:

- Escoger un patrón en las actividades de **Patrones de perímetro** (página 33) y **Patrones de superficie** (página 28) y ver si pueden ingresar una función constante para duplicar el patrón en ese cuadro de pares ordenados.

- Escoger uno de los cuadros de pares ordenados y ver si pueden pensar en una situación real que pudiera representar.

Name:

The Mysterious Constant
Recording Sheet

Collecting and Organizing Data

Number of times Cons has been pressed (left side of the display)	Result (right side of the display)

Analyzing Data and Drawing Conclusions

- A pattern we discovered in our table is:

- I think the constant function my partner entered on the calculator is _____ .

- The 95th time Cons is pressed, the number display will be_____ . We think this because:

Questions we thought of while we were doing this activity:

Nombre:

La misteriosa constante
Hoja de registro

Cómo reunir y organizar los datos

Número de veces que se presiona Cons (lado izquierdo del visor)	Resultado (lado derecho del visor)

Cómo analizar los datos y sacar conclusiones

- Describa un patrón que haya descubierto en su cuadro.

- Creo que la función constante que mi compañero ingresó en la calculadora es _____ .

- La 95ª vez que se presiona Cons , el número desplegado será _____. Así lo creemos porque:

Preguntas que surgieron mientras realizábamos esta actividad:

CONSTANT-ly

Math Concepts

- whole numbers
- comparing numbers
- estimation
- addition
- multiplication
- functions

Materials

- Math Explorer™
- **CONSTANT-ly** recording sheets
- pencils

Overview

Students will investigate a pattern using the constant function on the calculator, record the results displayed on the calculator, and describe how the constant key works.

Introduction

> **The Mysterious Constant** activity on page 38 should be completed before beginning this activity.

1. Using a transparency of the recording sheet, model this activity on the overhead projector. Have students choose a constant function and number. Write them in the box at the top left. Then have students enter the constant in their calculators.

   Example: For the constant + 3, have students enter ⊞ **3** Cons to prepare their calculators.

2. Ask students to choose the starting number for the pattern, 1, for example. Record it in the "Start with" box below the constant function. Then have students enter the starting number in their calculators.

3. Next, have students use their experiences with **The Mysterious Constant** (page 38) to predict what will happen each time they press Cons.

4. Test their conjectures. Choose a student to write both the constant function (constant operation and constant number) and the result (number on the right of the calculator display) on the transparency each time the rest of the students press Cons.

   Note: The counter (the number on the left of the calculator display) is already on the recording sheet.

CONSTANTE-mente

Conceptos matemáticos

- números enteros
- comparación numérica
- estimación

- suma
- mutiplicación
- funciones

Materiales

- Math Explorer™
- hojas de registro de **CONSTANTE-mente**
- lápices

Resumen

Los alumnos investigarán un patrón usando la función constante en la calculadora, registrarán los resultados que se ven en el visor y describirán cómo funciona la tecla constante.

Introducción

> Antes de comenzar con esta actividad, se debe completar el ejercicio de **La misteriosa constante** de la página 38.

1. Represente esta actividad en el proyector con una transparencia de la hoja de registro. Que los alumnos escojan un número y función constantes. Escríbalos en la casilla arriba a la izquierda. Que los alumnos ingresen la constante en la calculadora.

   Ejemplo: para la constante + 3, que los alumnos ingresen ⊞ **3** ⟦Cons⟧ para preparar la calculadora.

2. Pida a los alumnos que escojan el número inicial para el patrón. Por ejemplo, 1. Regístrelo en la casilla "Empezar con", debajo de la función constante. Después, que los alumnos ingresen el número inicial en la calculadora.

3. En seguida, que los alumnos apliquen su experiencia con **La misteriosa constante** (página 38) para predecir lo que sucederá cada vez que presionan ⟦Cons⟧.

4. Pruebe sus conjeturas. Escoja un alumno para que escriba en la transparencia la función constante (operación constante y número constante) y el resultado (número a la derecha en el visor de la calculadora), cada vez que el resto de los alumnos presionan ⟦Cons⟧.

   Nota: el contador (número a la izquierda en el visor de la calculadora) ya está en la hoja de registro.

CONSTANT-ly (continued)

Introduction (continued)

Example:

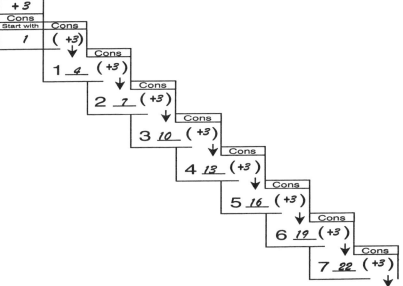

5. After they have pressed ⌈Cons⌉ seven times, have students describe what is happening each time, as well as any other patterns they notice.

6. Have students predict the next several numbers in the pattern and then verify their predictions with the calculator.

7. Have students work in pairs to choose a different constant function and starting number. Have one partner operate the calculator. Each time he or she presses ⌈Cons⌉, have the other partner record the constant function and the result that appears in the calculator display.

8. After exploring several different constant functions and starting numbers, have students explain how ⌈Cons⌉ works and describe a real-life situation where the constant function might be helpful.

Collecting and Organizing Data

While students explore their patterns, ask questions such as:

- What do the numbers on the left of your display represent? How about the numbers on the right of your display?

 How could you describe what the calculator is doing each time you press ⌈Cons⌉? What should come next if you are correct?

CONSTANTE-mente (continuación)

Introducción (continuación)

Ejemplo:

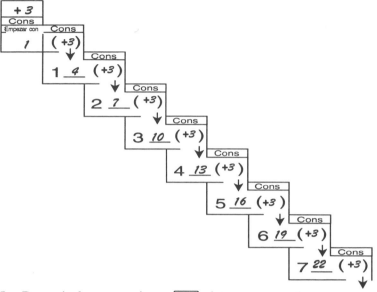

5. Después de que presionen [Cons] siete veces, que los alumnos describan lo que sucede cada vez y todos los patrones que observan.

6. Que los alumnos predigan los números siguientes en el patrón y que luego verifiquen sus predicciones con la calculadora.

7. Trabajando en parejas, que los alumnos escojan una función constante y un número inicial diferentes. Que un alumno ocupe la calculadora. Cada vez que presione [Cons], que el otro alumno registre en su hoja de registro la función constante y el resultado que se ve en el visor de la calculadora.

8. Después de probar varias funciones constantes y números iniciales diferentes, que los alumnos describan cómo funciona [Cons] y presenten una situación real en la que sería útil la función constante.

Cómo reunir y organizar los datos

Mientras los alumnos prueban sus patrones, haga las preguntas siguientes:

- ¿Qué representan los números en la parte izquierda del visor? ¿Y los números a la derecha?

¿Cómo podrían describir lo que hace la calculadora cada vez que presionan [Cons]? ¿Qué debería pasar después si tienen razón?

CONSTANT-ly (continued)

Collecting and Organizing Data (continued)

- Are the numbers on your recording sheet getting larger or smaller? By how much?

- What patterns do you notice?

- What would happen if the constant function remained the same but the starting number were different? How would the numbers on your recording sheet change?

 How can you use the calculator to predict the 95th number in your pattern?

Analyzing Data and Drawing Conclusions

After students have investigated several different constant functions and beginning numbers, have them work as a whole group to analyze the patterns. Ask questions such as:

 How did you use your calculator to help you make predictions?

- What are some strategies you used to predict how pressing [Cons] would affect your display?

- Describe the patterns you discovered.

- What happened when you kept the same constant function but started with different numbers?

- What did you notice when you kept the same beginning number but entered a different constant function?

- What did you write to explain how the constant key works?

- How could you use mathematical symbols to describe how [Cons] works?

- When might the constant function be useful?

Continuing the Investigation

Have students invent a constant function that involves more than one operation. Ask them: Can you predict the pattern that would develop if you entered your beginning number and kept pressing [Cons]? How could you check your prediction?

CONSTANTE-mente (continuación)

Cómo reunir y organizar los datos (continuación)

- ¿Los números en su hoja de registro se hacen más grandes o más pequeños? ¿En qué proporción?

- ¿Qué patrones observan?

- ¿Qué sucedería si la función constante siguiera siendo la misma, pero el número inicial fuera diferente? ¿Cómo cambiarían los números en su hoja de registro?

⊞ ¿Cómo pueden usar la calculadora para predecir el 95º número en su patrón?

Cómo analizar los datos y sacar conclusiones

Después de que los alumnos investiguen varias funciones constantes y números iniciales diferentes, pídales que analicen los patrones. Haga las preguntas siguientes:

- ¿Qué estrategias usaron para predecir el resultado al presionar [Cons]?

- Describan los patrones que descubrieron

- ¿Qué sucedió cuando mantuvieron la misma función constante, pero comenzaron con números diferentes?

- ¿Qué observaron cuando mantuvieron el mismo número inicial, pero ingresaron una función constante diferente.

- ¿Qué escribieron sobre la forma en que funciona la tecla constante?

- ¿Cómo podrían utilizar símbolos matemáticos para describir cómo funciona [Cons]?

- ¿Cúando sería útil la función constante?

⊞ ¿Cómo usaron la calculadora para hacer predicciones?

⊞ ¿Cómo usaron la calculadora para descubrir los patrones en los pares ordenados de sus cuadros?

Cómo continuar la investigación

Que los alumnos inventen una función constante que involucre más de una operación. Que predigan el patrón que se desarrollaría si ingresaran su número inicial y siguieran presionando la tecla [Cons]. ¿Cómo comprobarían su predicción?

Name:

CONSTANT-ly
Recording Sheet

Collecting and Organizing Data

How Cons works:

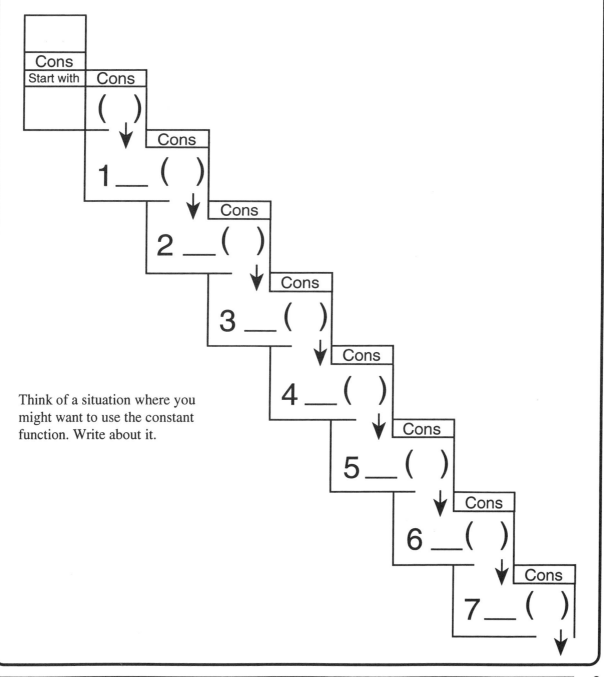

Think of a situation where you
might want to use the constant
function. Write about it.

Nombre:

CONSTANTE-mente
Hoja de registro

Cómo reunir y organizar los datos

La tecla [Cons] funciona así:

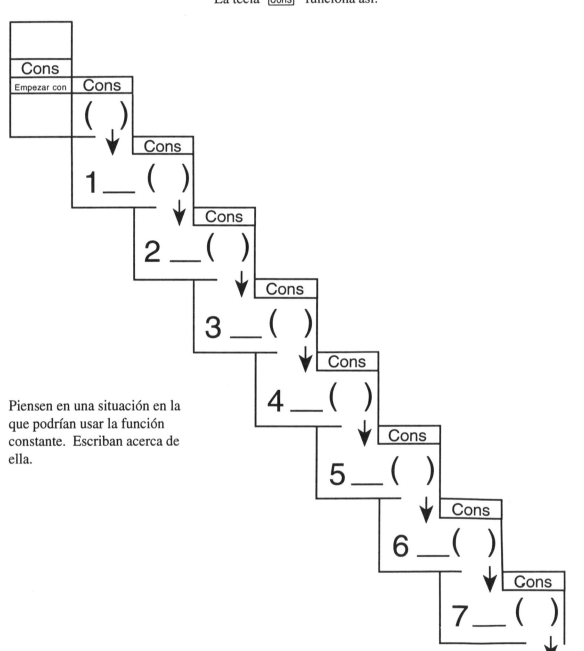

Piensen en una situación en la que podrían usar la función constante. Escriban acerca de ella.

"Power"ful Patterns

Math Concepts

- multiplication
- division
- exponents
- place value
- whole numbers
- decimals

Materials

- Math Explorer™
- **"Power"ful Patterns** recording sheets
- pencils

Overview

Students will investigate the relationship between multiplication with repeated factors and the use of exponents. Students will connect "powers of ten" to place value positions.

Introduction

1. Have students investigate a problem that requires multiplication of a repeated factor.

   Examples:

   - If you fold a piece of paper in half five times, how many layers thick would it be?

   - If you didn't like baby-sitting for a particular child, you might decide to triple your hourly charge each time you baby-sit. What would you charge on the fourth time, if you charged $3.00 an hour the first time?

   - After a certain kind of plant is planted, it produces ten seeds a season. Then it stops producing seeds. If each of those ten seeds grows into a plant with the same seed-producing cycle, how many seeds will you have at the end of six seasons?

2. Have students discuss how they might use the calculator to perform the repeated multiplication actions in the problem they choose to investigate.

   Examples:
   2 $\boxed{\times}$ 2 $\boxed{\times}$ 2 $\boxed{\times}$ ... $\boxed{=}$ _____

   or

   $\boxed{\times}$ 2 $\boxed{\text{Cons}}$ $\boxed{=}$ 1 $\boxed{\text{Cons}}$ $\boxed{\text{Cons}}$ $\boxed{\text{Cons}}$ ....

3. Demonstrate using the exponent key $\boxed{y^x}$ for multiplication of a repeated factor.

Patrones de potencias

Conceptos matemáticos

- multiplicación
- división
- exponentes
- valor de posición
- números enteros
- decimales

Materiales

- Math Explorer™
- hojas de registro de **Patrones de potencias**
- lápices

Resumen

Los alumnos investigarán la relación entre multiplicación con factores repetidos y el uso de exponentes. Relacionarán las "potencias decimales" con los valores de posición.

Introducción

1. Que los alumnos investiguen un problema que requiere la multiplicación de un factor repetido.

    Ejemplos:

    - Si doblan un pedazo de papel por la mitad cinco veces, ¿cuántas capas de grosor tendría al final?

    - Supongan que no les gusta hacer clases de matemáticas a un niño determinado; la solución podría ser triplicar sus honorarios por hora cada vez que le hacen clases a ese niño. ¿Cuánto cobrarían la cuarta vez si la primera cobraron $3.00 por hora?

    - Después de plantar un cierto tipo de planta, ésta produce diez semillas una temporada y después deja de producir. Si cada una de esas diez semillas se transforma en una planta con el mismo ciclo de producción de semillas, ¿cuántas semillas tendrían después de seis temporadas?

2. Que los alumnos discutan cómo podrían utilizar la calculadora para ejecutar acciones de multiplicación repetidas en el problema que decidieron investigar.

    Ejemplos:
    2 ⨯ 2 ⨯ 2 ⨯ ... ⊟ ____

    o

    ⨯ 2 Cons ⊟ 1 Cons Cons Cons ....

3. Demuestre usando la tecla exponecial y^x para multiplicación de un factor repetido.

"Power"ful Patterns (continued)

Introduction (continued)

4. Have students use both $\boxed{y^x}$ and $\boxed{\text{Cons}}$ to generate data about the problem. Have them record the data in the tables on the recording sheet.

   Note: In exponential notation, the repeated factor is called the "base," and the exponent is often referred to as the "power."

   Example:

Repeated Multiplication with $\boxed{y^x}$			Repeated Multiplication with $\boxed{\text{Cons}}$		
Enter: (base) $\boxed{y^x}$ (exponent) $\boxed{=}$			Enter: × *3* $\boxed{\text{Cons}}$ 1 (base)		
y(base)	*x*(exponent)	product			product
3 $\boxed{y^x}$ *1*	=	*3*	$\boxed{\text{Cons}}$	1	*3*
3 $\boxed{y^x}$ *2*	=	*9*	$\boxed{\text{Cons}}$	2	*9*
3 $\boxed{y^x}$ *3*	=	*27*	$\boxed{\text{Cons}}$	3	*27*
3 $\boxed{y^x}$ *4*	=	*81*	$\boxed{\text{Cons}}$	4	*81*
3 $\boxed{y^x}$ *5*	=	*243*	$\boxed{\text{Cons}}$	5	*243*
3 $\boxed{y^x}$ *6*	=	*729*	$\boxed{\text{Cons}}$	6	*729*

5. Have students compare the data in the tables and discuss the relationships between the two lists.

Collecting and Organizing Data

While students generate data for the different bases (factors) and exponents (powers) in their problems, ask questions such as:

- What factor are you using? How is it represented in your problem?

- What does the exponent represent?

- Predict what the next entry in your table should be. How did you make your prediction?

- What happens when you change the exponent?

- What happens when you change the factor (the base of the exponent)?

 How are the different ways to perform repeated multiplication on the calculator alike? How are they different?

 What happens if you enter 0 instead of 1 as the starting number for the constant function? Why do you think this happens?

Patrones de potencias (continuación)

Introducción (continuación)

4. Que los alumnos utilicen las teclas $\boxed{y^x}$ y $\boxed{\text{Cons}}$ para generar datos sobre el problema. Que anoten los datos en los cuadros de la hoja de registro.

Nota: en notación exponencial, el factor repetido se denomina "base" y el exponente se conoce comúnmente como "potencia".

Ejemplo:

Multiplicación repetida con $\boxed{y^x}$						Multiplicación repetida con $\boxed{\text{Cons}}$		
Ingresar: (base) $\boxed{y^x}$ (exponente) $\boxed{=}$						Ingresar: ×3 $\boxed{\text{Cons}}$ 1 (base)		
y(base)		x(exponente)		producto				producto
3	$\boxed{y^x}$	1	=	3		$\boxed{\text{Cons}}$	1	3
3	$\boxed{y^x}$	2	=	9		$\boxed{\text{Cons}}$	2	9
3	$\boxed{y^x}$	3	=	27		$\boxed{\text{Cons}}$	3	27
3	$\boxed{y^x}$	4	=	81		$\boxed{\text{Cons}}$	4	81
3	$\boxed{y^x}$	5	=	243		$\boxed{\text{Cons}}$	5	243
3	$\boxed{y^x}$	6	=	729		$\boxed{\text{Cons}}$	6	729

5. Que los alumnos comparen los datos de los cuadros y analicen las relaciones entre las dos listas.

Cómo reunir y organizar los datos

Mientras los alumnos generan datos para los diferentes exponentes (potencias) y bases (factores) de sus problemas, haga las preguntas siguientes:

- ¿Qué factor están usando? ¿Cómo está representado en su problema?

- ¿Qué representa el exponente?

- Pronostiquen cuál será la próxima entrada en su cuadro. ¿Cómo hicieron su predicción?

- ¿Qué sucede cuando cambian el exponente?

- ¿Qué sucede cuando cambian el factor (la base del exponente)?

¿En qué se parecen las distintas formas de ejecutar una multiplicación repetida en la calculadora? ¿En qué se diferencian?

¿Qué sucede si ingresan 0 en vez de 1 como número inicial para la función constante? ¿Por qué piensan que ocurre esto?

"Power"ful Patterns (continued)

Analyzing Data and Drawing Conclusions

After students have made and compared several pairs of lists using different bases (factors), have them discuss their results as a whole group. Ask questions such as:

- What problem did you make up to generate your data?

- How are everyone's problems alike? How are they different?

- How are the different data lists alike? How are they different?

- What kinds of relationships do you see between the two lists of data in each chart?

- What does y^x represent? How else could you represent that idea with the calculator?

- What kinds of products occurred when you used 10 as the factor (the base)? What patterns did you see in them? Where else have you seen that pattern?

When would you most likely want to use the regular multiplication key instead of y^x or [Cons]? When would y^x or [Cons] be helpful?

Continuing the Investigation

Have students:

- Experiment with using negative numbers for the factor (base) or for the exponent.

- Experiment with using division instead of multiplication as the repeated operation.

- Look for relationships between the data collected in these experiments.

Patrones de potencias (continuación)

Cómo analizar los datos y sacar conclusiones

Después de que los alumnos creen y comparen varios pares de listas usando bases (factores) diferentes, que analicen sus resultados como un solo grupo. Haga las preguntas siguientes:

- ¿Qué problema enfrentaron para generar sus datos?

- ¿En qué se parecen los problemas individuales? ¿En qué se diferencian?

- ¿En qué se parecen las distintas listas de datos? ¿En qué se diferencian?

- ¿Qué tipos de relaciones observan entre las dos listas de datos en cada cuadro?

- ¿Qué representa y^x? ¿De qué otra manera podrían representar esa idea con la calculadora?

- ¿Qué tipos de producto se dan cuando usan 10 como factor (base)? ¿Qué patrones observaron en ellos? ¿En qué otra parte vieron ese patrón?

¿Cuándo usarían con mayor probabilidad la tecla de multiplicación normal en lugar de [y^x] o [Cons]? ¿Cuándo serían útiles [y^x] o [Cons]?

Cómo continuar la investigación

Los alumnos deben:

- Experimentar con el uso de números negativos para el factor (base) o para el exponente.

- Experimentar con el uso de la división en lugar de la multiplicación como operación repetida.

- Buscar relaciones entre los datos reunidos en estos experimentos.

Name:

"Power"ful Patterns

Recording Sheet

Collecting and Organizing Data

Repeated Multiplication with $\boxed{y^x}$				Repeated Multiplication with $\boxed{\text{Cons}}$		
Enter: (base) $\boxed{y^x}$ (exp) $\boxed{=}$				Enter: $\underline{\times}\underline{}$ $\boxed{\text{Cons}}$ **1**		
y (base)	x (exponent)		product			product
_____	$\boxed{y^x}$ 1	=	_____	$\boxed{\text{Cons}}$ 1		_____
_____	$\boxed{y^x}$ 2	=	_____	$\boxed{\text{Cons}}$ 2		_____
_____	$\boxed{y^x}$ 3	=	_____	$\boxed{\text{Cons}}$ 3		_____
_____	$\boxed{y^x}$ 4	=	_____	$\boxed{\text{Cons}}$ 4		_____
_____	$\boxed{y^x}$ 5	=	_____	$\boxed{\text{Cons}}$ 5		_____
_____	$\boxed{y^x}$ 6	=	_____	$\boxed{\text{Cons}}$ 6		_____
_____	$\boxed{y^x}$ 7	=	_____	$\boxed{\text{Cons}}$ 7		_____
_____	$\boxed{y^x}$ 8	=	_____	$\boxed{\text{Cons}}$ 8		_____
_____	$\boxed{y^x}$ 9	=	_____	$\boxed{\text{Cons}}$ 9		_____

Analyzing Data and Drawing Conclusions

- What is the calculator doing when you enter: $\underline{}_{\text{(base)}}$ $\boxed{y^x}$ $\underline{}_{\text{(exponent)}}$ = ?

- What is the calculator doing each time you press $\boxed{y^x}$?

- What pattern do you see when you use 10 as the base?

Nombre:

Patrones de potencias
Hoja de registro

Cómo reunir y organizar los datos

Multiplicación repetida con y^x Ingresar: (base) y^x (exponente) $=$				Multiplicación repetida con Cons Ingresar: $\times \underline{\quad}$ Cons **1** (base)	
y (base)	**x (exponente)**		**producto**		**producto**
_____	y^x 1	=	_____	Cons 1	_____
_____	y^x 2	=	_____	Cons 2	_____
_____	y^x 3	=	_____	Cons 3	_____
_____	y^x 4	=	_____	Cons 4	_____
_____	y^x 5	=	_____	Cons 5	_____
_____	y^x 6	=	_____	Cons 6	_____
_____	y^x 7	=	_____	Cons 7	_____
_____	y^x 8	=	_____	Cons 8	_____
_____	y^x 9	=	_____	Cons 9	_____
_____	y^x 10	=	_____	Cons 10	_____

Cómo analizar los datos y sacar conclusiones

- ¿Qué hace la calculadora cuando ingresan: _____ y^x _____ = ?
  (base)   (exponente)

- ¿Qué hace la calculadora cada vez que presionan y^x ?

- ¿Qué patrones observan cuando utilizan 10 como base?

Patterns in Division

Math Concepts

- division
- whole numbers
- fractions
- decimals

Materials

- Math Explorer™
- **Patterns in Division** recording sheets
- objects for counters
- pencils

Overview

Students will investigate how the quotients and remainders from integer division relate to the quotients from rational number division in both fraction and decimal form.

Introduction

1. Pose a simple problem such as: If you had ten cookies to divide evenly among three people, how many cookies would each person get? Have students act out the division with counters. (Or, use the cookie pictures provided on page 54.)

   Ask: What would you do with the left-over cookie? Discuss the two choices: Leave the cookie alone and let it be left over, or cut it into three equal pieces to give to the three people.

2. Discuss how you might record the first choice with a whole number quotient and a whole number remainder and how you might record the second choice with a fraction in the quotient.

3. Demonstrate using the integer divide key [INT÷] on the calculator to record the situation where you leave the extra cookie alone: **10** [INT÷] **3** [=] to display ⌊ ₍ₒ₎3 ⌋ ⌊ ₍ᵣ₎1 ⌋.

4. Have students compare this method of division to dividing in fraction notation, displaying the quotient in mixed number form. Enter **10** [/] **3** to represent 10 divided by 3 as a fraction. Then press [Ab/c] to display 3⌐1/3, the mixed number quotient for 10 divided by 3.

5. Finally, have students change the fraction to decimal form using [F⊃D].

6. Have students compare this result to the decimal quotient found by entering **10** [÷] **3** [=].

7. Have students record their data on their recording sheets.

Patrones de la división

Conceptos matemáticos

- división
- fracciones
- números enteros
- decimales

Materiales

- Math Explorer™
- hojas de registro de **Patrones de la división**
- objetos para usar como contadores
- lápices

Resumen

Los alumnos investigarán cómo se relacionan los cocientes y los restos de la división de enteros con los cocientes de la división de números racionales en forma de fracción y decimal.

Introducción

1. Plantee un problema simple como el siguiente: si tuvieran diez galletas para repartir en forma equitativa entre tres personas, ¿cuántas galletas tendría cada persona? Que los alumnos realicen la división con contadores (o que usen los dibujos de galletas que vienen en la página 54).

   Pregunta: ¿qué harían con la galleta que sobra? Analicen las dos alternativas: no repartirla o cortarla en tres partes iguales para repartirla entre las tres personas?

2. Analice la forma en que podrían registrar la primera alternativa con un cociente y un resto en números enteros, y la segunda alternativa, con una fracción en el cociente.

3. Demuestre usando la tecla de división de enteros $\boxed{\text{INT}\div}$ de la calculadora, para registrar la situación en la que no se reparte la galleta que sobra: **10** $\boxed{\text{INT}\div}$ **3** $\boxed{=}$ para desplegar $\lfloor_Q 3 \rfloor \lfloor_R 1 \rfloor$.

4. Que los alumnos comparen este método de división con el de notación de fracción, desplegando el cociente en forma de número mixto. Ingrese **10** $\boxed{/}$ **3** para representar 10 dividido por 3 como una fracción. Después, presione $\boxed{\text{Ab/c}}$ para desplegar **3␣1/3**, el cociente en número mixto para 10 dividido por 3.

5. Finalmente, que los alumnos cambien la fracción a forma decimal usando $\boxed{\text{F⊃D}}$.

6. Que los alumnos comparen este resultado con el cociente decimal que resultó al ingresar **10** $\boxed{\div}$ **3** $\boxed{=}$.

7. Que los alumnos anoten sus datos en las hojas de registro.

Patterns in Division (continued)

Introduction (continued)

8. Ask students: What data do you think you would collect if you had started with a different number of cookies? What if there had been 20 cookies instead of 10? 30 cookies? 35 cookies?

9. Have students work in small groups to investigate what happens when they use four as a divisor. Have them record the data on their recording sheets, try other divisors, and look for patterns in the data.

Collecting and Organizing Data

While students generate data for the different divisors, ask questions such as:

- What divisor are you using? What kinds of remainders do you expect to find? (Refer to the **Recurring Remainders** activity on page 10.)

- What does the quotient represent? What does the remainder represent?

- What does the fraction form of the quotient represent?

- What patterns do you notice in the fraction form of the quotients? How do the fractions relate to the remainders? To the divisors? Why do you think this happens?

- What happens when you change the divisor? How do your remainders and fractions change? How do their relationships change?

- What relationships do you see between the fractions and the decimals?

- What kind of conjectures can you make about relationships between the remainders and the fractions? The relationships between the fractions and the decimals?

How are the different ways to perform division on the calculator alike? How are they different?

Patrones de la división (continuación)

Introducción (continuación)

8. Pregunta para los alumnos: ¿qué datos piensan que reunirían si empezaran con un número distinto de galletas? ¿Qué sucedería si hubiera 20 galletas en vez de diez? ¿Y con 30 ó 35?

9. Que los alumnos trabajen en grupos pequeños para investigar lo que sucede cuando utilizan cuatro como divisor. Pídales que anoten los datos en sus hojas de registro, prueben con otros divisores y busquen patrones en los datos.

Cómo reunir y organizar los datos

Mientras los alumnos generan datos para los diferentes divisores, haga las preguntas siguientes:

- ¿Qué divisor están usando? ¿Qué tipos de resto esperan encontrar? (ver la actividad **Restos recurrentes** en la página 10).

- ¿Qué representa el cociente? ¿Qué significa el resto?

- ¿Qué representa la forma de fracción del cociente?

- ¿Qué patrones observaron en la forma de fracción de los cocientes? ¿Cómo se relacionan las fracciones con los restos? ¿Con los divisores? ¿Por qué creen que sucede esto?

- ¿Qué sucede cuando cambian el divisor? ¿Cómo cambian los restos y las fracciones? ¿Cómo cambian sus relaciones?

- ¿Qué relaciones observan entre las fracciones y los decimales?

- ¿Qué tipo de conjeturas pueden formular acerca de las relaciones entre los restos y las fracciones? ¿Y sobre las relaciones entre las fracciones y los decimales?

¿En qué se parecen las distintas formas de ejecutar una división en la calculadora? ¿En qué se diferencian?

Patterns in Division (continued)

Analyzing Data and Drawing Conclusions

After students have made and compared several tables and looked for relationships, have them discuss their results as a whole group. Ask questions such as:

- What kinds of remainders occur with a divisor of ____? Why?

- What kinds of fractions occur in the quotients with a divisor of ____? Why?

- What kinds of decimals occur in the quotients with a divisor of ____? Why do you think those decimals look the way they do?

- Do the related fraction quotients and decimal quotients represent the same amount? Why do you think so or not think so?

- Now what kinds of questions do you have about division, remainders, fractions, and decimals? How might we investigate these questions?

▦ When would you most likely want to use [INT÷]?

▦ When would you want to use [÷] on the calculator?

▦ When might you want to represent division as a fraction with [/] on the calculator?

Continuing the Investigation

Have students collect data about other divisors to see if their conjectures continue to be true or if the patterns in their data lead them to new conjectures.

Patrones de la división (continuación)

Cómo analizar los datos y sacar conclusiones

Después de que los alumnos creen y comparen varios cuadros y busquen relaciones, pídales que analicen sus resultados como un solo grupo. Haga las preguntas siguientes:

- ¿Qué tipos de resto se dan con un divisor de _____? ¿Por qué?

- ¿Qué tipos de fracciones se dan en los cocientes con un divisor de _____? ¿Por qué?

- ¿Qué tipos de decimales se dan en los cocientes con un divisor de _____? ¿Por qué piensan que esos decimales son así?

- ¿Los cocientes de fracción y los cocientes decimales relacionados representan la misma cantidad? Expliquen por qué piensan así.

- ¿Qué tipo de preguntas tienen ahora acerca de la división, restos, fracciones y decimales? ¿Cómo podríamos investigar estas preguntas?

¿Cuándo desearían con mayor probabilidad usar INT÷?

¿Cuándo desearían usar la tecla ÷ en la calculadora?

¿Cuándo desearían representar la división como una fracción con la tecla / en la calculadora?

Cómo continuar la investigación

Que los alumnos reúnan datos acerca de otros divisores para ver si mantienen sus conjeturas o si los patrones en sus datos los llevan a otras nuevas.

Name:

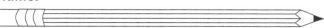

Patterns in Division
Recording Sheet

Collecting and Organizing Data

# of Cookies (Dividend)	# of People (Divisor)	Integer Quotient & Remainder	Fraction Quotient	Decimal Quotient

Analyzing Data and Drawing Conclusions

Questions we thought of while we were doing this activity:

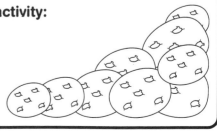

Nombre:

Patrones de la división
Hoja de registro

Cómo reunir y organizar los datos

# de galletas (Dividendo)	# de personas (divisor)	Cociente y resto en enteros	Cociente en fracción	Cociente Decimal

Cómo analizar los datos y sacar conclusiones

Preguntas que surgieron mientras realizábamos esta actividad:

Patterns in Division

Cookies

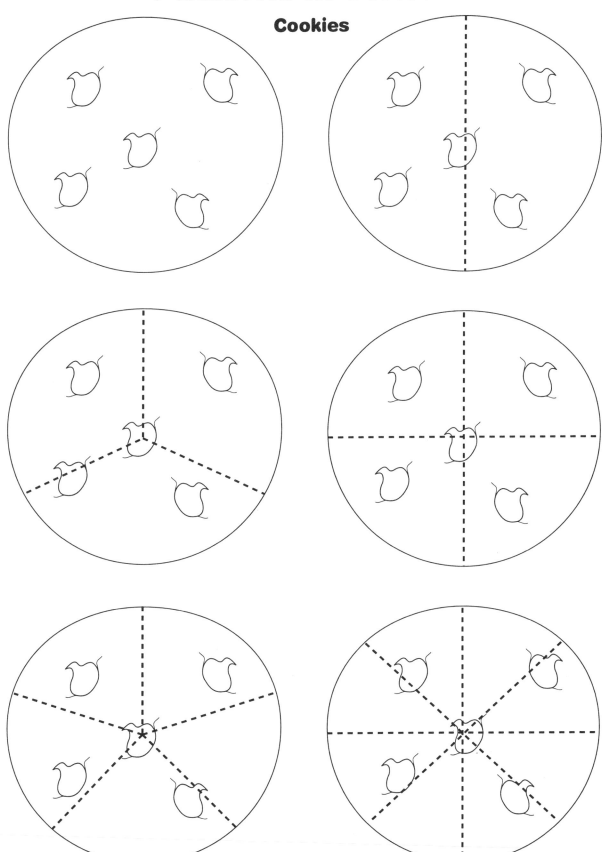

Patrones de la división
Galletas

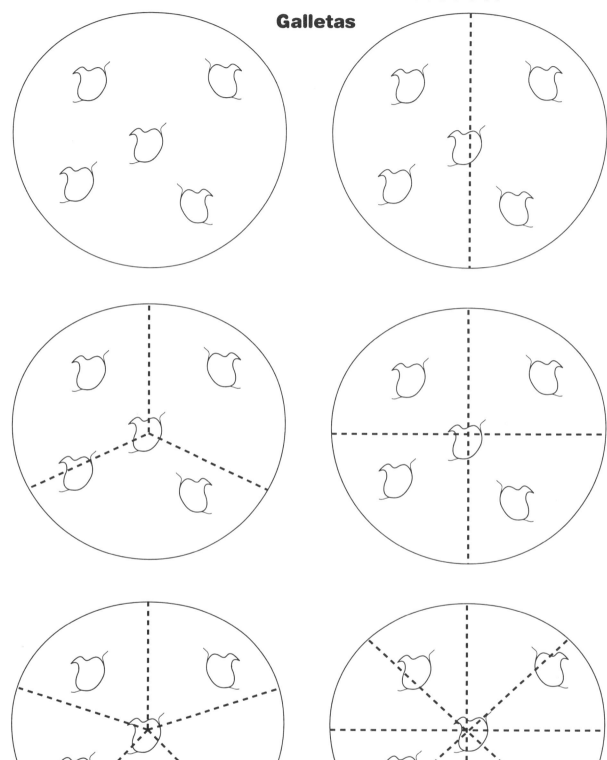

Picturing Percents

Math Concepts

- multiplication
- division
- place value
- whole numbers
- fractions
- decimal
- ratio
- proportion
- percent

Materials

- Math Explorer™
- **Picturing Percents** recording sheets
- pencils

Overview

Students will represent percents on a 10 x 10 grid. They will use the grid and the calculator to generate patterns that lead to methods for calculating percentages.

Introduction

1. Discuss with students the origins of the word *percent* — *per* means "for each" and *cent* means "hundred."

2. Show a 10 × 10 grid (page 59) on the overhead projector, and ask questions such as:

   - How many squares are in each row?
   - How many squares are in each column?
   - How many small squares are in the entire large square (grid)?
   - If the large square (grid) represents 1, what does each small square represent?
   - If the large square (grid) is described as 100%, what percent does each small square represent?

3. Have students practice representing various percentages on the 10 × 10 grid.

   Example: For 25%, color 25 small squares in some configuration.

4. Introduce a problem such as: Jorge makes a 12% commission on the newspapers he sells. If he sells $100 worth of papers, how much money in commission does he make?

   Have students model a solution to this problem on the 10 × 10 grid.

   Example: If the entire grid represents 100% of the $100 total, one small square represents 1% (1/100) of the total, or $1. Then 12 small squares represents:

   12% (12/100) of the total, or 12 × $1, or $12 commission.

Representación de los porcentajes

MATH
EXPLORER

Conceptos matemáticos

- multiplicación
- división
- valor de posición
- números enteros
- fracciones
- decimales
- relación
- proporción
- porcentaje

Materiales

- Math Explorer™
- hojas de registro de **Representación de los porcentajes**
- lápices

Resumen

Los alumnos representarán porcentajes en una cuadrícula de 10 × 10. Usarán la cuadrícula y la calculadora para generar patrones que lleven a métodos para calcular porcentajes.

Introducción

1. Analice con los alumnos el significado de la palabra porcentaje.

2. Muestre en el proyector de transparencias una cuadrícula de 10 × 10 y haga las preguntas siguientes:

    - ¿Cuántos cuadrados hay en cada fila?

    - ¿Cuántos cuadrados hay en cada columna?

    - ¿Cuántos cuadrados hay en total en la cuadrícula?

    - Si la cuadrícula completa representa 1, ¿qué representa cada cuadrado que la compone?

    - Si la cuadrícula completa se describe como 100%, ¿qué porcentaje representa cada cuadrado que la compone?

3. Que los alumnos practiquen la representación de varios porcentajes en la cuadrícula de 10 × 10.

    Ejemplo: para 25%, que coloreen 25 cuadrados en alguna configuración determinada.

4. Plantee un problema como el siguiente: Jorge obtiene 12% de comisión por la venta de periódicos. Si vende $100 en periódicos, ¿cuánto dinero gana como comisión?

    Que los alumnos representen una solución a este problema en la cuadrícula de 10 × 10.

    Ejemplo: si la cuadrícula entera representa 100% de los $100 totales, un cuadrado representa 1% (1/100) del total o $1. Por lo tanto, 12 cuadrados representan:

    12% (12/100) del total, o 12 × $1 o $12 de comisión.

Picturing Percents (continued)

Introduction (continued)

5. Discuss with students the percent problem on the recording sheet. Have them fill in blanks (a) and (b) and model solutions for blank (c) using the 10×10 grid. Have them use the table to record their data for the solution to the problem.

6. Have students work in small groups to generate several different problems, record their data, and look for patterns to develop a method for calculating percentages.

7. Have students use their calculators to test methods for calculating percentages.

Collecting and Organizing Data

While students generate data from the solutions to their different problems, ask questions such as:

*   What does the 10×10 grid represent in general? What does it represent in this particular problem?

*   What does each small square represent in general? What does each small square represent in this particular problem?

*   How did you go about finding the value of a small square?

*   Predict the solution to your problem. How did you make your prediction?

*   How can you use the value of a small square to help you find the solution to your problem?

*   What patterns do you see in the table information used to find your solutions?

How can the calculator be used to help you determine the value of each small square?

How can the calculator be used to help you determine the solution to the problem?

Representación de los porcentajes

(continuación)

Introducción (continuación)

5. Analice con los alumnos el problema de porcentaje de la hoja de registro. Que llenen los espacios en blanco (a) y (b) y representen soluciones para el espacio (c) usando la cuadrícula de 10×10. Que utilicen el cuadro para registrar sus datos para la solución del problema.

6. Que los alumnos trabajen en grupos pequeños para generar varios problemas diferentes, registren sus datos y busquen patrones para desarrollar un método para calcular porcentajes.

7. Que los alumnos utilicen la calculadora para probar los métodos para calcular porcentajes.

Cómo reunir y organizar los datos

Mientras los alumnos generan datos a partir de las soluciones de los distintos problemas, haga las preguntas siguientes:

- ¿Qué representa, en general, la cuadrícula de 10×10? ¿Qué representa en este problema específico?

- ¿Qué representa, en general, cada cuadrado pequeño? ¿Qué representa cada cuadrado pequeño en este problema específico?

- ¿Cómo procedieron para encontrar el valor de un cuadrado pequeño?

- Pronostiquen la solución de su problema. ¿Cómo hicieron su predicción?

- ¿Cómo pueden usar el valor de un cuadrado pequeño para encontrar la solución de su problema?

- ¿Qué patrones observan en la información tabular usada para encontrar sus soluciones?

¿Cómo pueden utilizar la calculadora para determinar el valor de cada cuadrado pequeño?

¿Cómo pueden utilizar la calculadora para determinar la solución del problema?

Picturing Percents (continued)

Analyzing Data and Drawing Conclusions

After students have generated data and solutions for several different problems, have them discuss their results as a whole group. Ask questions such as:

- What numbers did you find most interesting in the problems you made up? Why?

- How are your data and solutions like everyone else's? How are they different?

- What relationships do you see among the three columns of the table, if any?

- How did you find the value of a small square? Did you use the same procedure for each problem?

- What generalizations could you make about the relationship of 1% of a number to the whole number (100%)?

- Using the data in your table, what other generalizations could you make about finding percentages?

Continuing the Investigation

Have students investigate percentages that are more than 100% and less than 1%. Do the same relationships in the data hold? Do the same procedures hold?

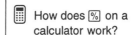 If your calculator has a %️ key, think about the following questions:

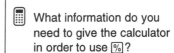 How does %️ on a calculator work?

What information do you need to give the calculator in order to use %️?

 What information does the calculator provide you when you use %️?

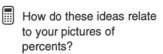 How do these ideas relate to your pictures of percents?

Representación de los porcentajes

(continuación)

Cómo analizar los datos y sacar conclusiones

Después de que los alumnos generen datos y soluciones para varios problemas distintos, que analicen sus resultados como un solo grupo. Haga las preguntas siguientes:

- ¿Qué números encontraron más interesantes en los problemas que resolvieron? ¿Por qué?

- ¿En qué se parecen sus datos y soluciones con los de los demás? ¿En qué se diferencian?

- ¿Qué relaciones, si las hay, observan entre las tres columnas del cuadro?

- ¿Cómo encontraron el valor de un cuadrado? ¿Usaron el mismo procedimiento para todos los problemas?

- ¿Qué afirmaciones generales se pueden formular acerca de la relación de 1% de un número con el número entero (100%)?

- Con los datos de su cuadro, ¿qué otras generalizaciones podrían formular acerca de cómo encontrar porcentajes?

Si la calculadora tiene una tecla %, piensen en las preguntas siguientes:

- ¿Cómo funciona % en la calculadora?

- ¿Qué información necesitan darle a la calculadora para usar %?

- ¿Qué información les entrega la calculadora cuando usan %?

- ¿Cómo se relacionan estas ideas con sus representaciones de los porcentajes?

Cómo continuar la investigación

Que los alumnos investiguen porcentajes superiores a 100% e inferiores a 1%. ¿Se mantienen las mismas relaciones en los datos? ¿Y los mismos procedimientos?

Name: _____

Picturing Percents
Recording Sheet

Collecting and Organizing Data

Using different sets of numbers in blanks (a) and (b) to determine the number in blank (c), fill in the table below and look for patterns.

Problem: Susan planted (a) _____ bulbs. The nursery guarantees that at least (b) _____ % of them will bloom within two weeks. (c) _____ bulbs must bloom for the guarantee to be upheld.

100% 10 × 10 Grid Represents Total Number of Bulbs	1% A Small Square Represents How Many Bulbs?	b Equals What Percent?	The Number of Blooming Bulbs in b%
a =		_____ %	c =
a =		_____ %	c =
a =		_____ %	c =
a =		_____ %	c =

Analyzing Data and Drawing Conclusions

From the patterns in the data above, we think we would find 15% of 360 by:

Nombre:

Representación de los porcentajes
Hoja de registro

Cómo reunir y organizar los datos

Llenen el cuadro siguiente y busquen patrones, utilizando distintos grupos de números en los espacios en blanco (a) y (b) para determinar el número en el espacio en blanco (c).

Problema: Susan plantó (a) _____ bulbos. El vivero garantiza que al menos (b) _____% de ellos brotarán en dos semanas. (c) _____ bulbos deben brotar para que se cumpla la garantía.

100% La cuadrícula 10 × 10 representa el número total de bulbos	1% ¿Cuántos bulbos representa un cuadrado?	b ¿Equivale a que porcentaje?	El número de bulbos que brotan en b%
a =		_____%	c =
a =		_____%	c =
a =		_____%	c =
a =		_____%	c =

Cómo analizar los datos y sacar conclusiones

A partir de los patrones en los datos anteriores, pensamos que encontraríamos 15% de 360 mediante el procedimiento siguiente:

Picturing Percents

10 x 10 Grid

Representación de los porcentajes

Cuadrícula de 10 x 10

Activities:

1 How Owl Flies

2 Map It!

3 Only Half There?

4 No More Peas, Please!

5 Do Centimeters Make Me Taller?

6 What's My Ratio?

7 Ratios in Regular Polygons

8 Predicting π

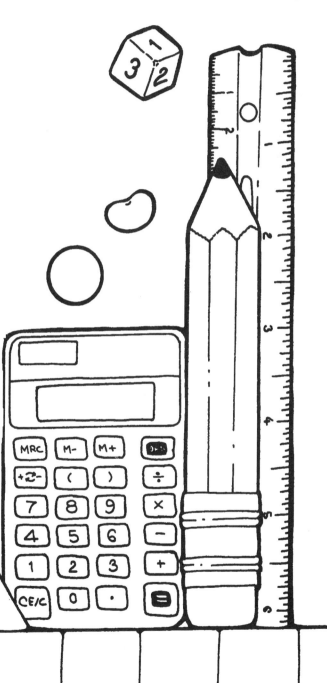

How Owl Flies

Math Concepts

- whole numbers
- fractions
- addition
- linear units of measure

Materials

- Math Explorer™
- map of 100 Aker Wood
- **How Owl Flies** recording sheets
- linear measuring tools (rulers, tape measures, string, etc.)
- pencils

Overview

Students will use the map legend from *Winnie the Pooh* to determine the distances that characters in the story must travel between different locations on the map.

Introduction

1. Use the map from the end papers of *Winnie the Pooh* by A. A. Milne to show the 100 Aker Wood. Project a transparency of the map (page 64) on an overhead screen. Choose two locations on the map and have students discuss the different ways that characters in the story could get from the first location to the second.

   Example: Owl would fly straight from one point to another while Tigger would bounce erratically from point to point to point . . . .

2. Distribute copies of the map to pairs of students. Have them use calculators and the map legend to determine the distance of Owl's straight line path from the first location to the second. Then have them determine the distances of other characters' paths along the ground from the first location to the second.

   Note: Students could use broken toothpicks to represent the paths.

3. Ask students to choose other pairs of locations on the map. Have them record the distances that Owl travels between the two locations and the distances traveled by each of the other characters in the story between the same two locations.

4. Ask students to analyze the recorded information and write about what they notice.

Cómo vuela el búho

Conceptos matemáticos

- números enteros
- fracciones
- suma
- unidades lineales de medida

Materiales

- Math Explorer™
- mapa de 100 Aker Wood
- hojas de registro de **Cómo vuela el búho**
- herramientas de medición lineal (reglas, cinta, cuerda, etc.)
- lápices

Resumen

Los alumnos utilizarán la escala del mapa de *Winnie the Pooh* para determinar las distancias que deben recorrer los personajes de la historia entre los distintos puntos del mapa.

Introducción

1. Use el mapa de las hojas finales de *Winnie the Pooh,* de A. A. Milne, donde aparece 100 Aker Wood. Proyecte una transparencia del mapa (página 64) en la pantalla del proyector. Escoja dos posiciones en el mapa y pida a los alumnos que analicen las distintas maneras en que los personajes de la historia podrían llegar desde el primer punto del mapa al segundo.

   Ejemplo: el búho volaría directamente desde un punto al otro, mientras que Tigger iría erráticamente entre un punto y otro. . . .

2. Distribuya copias del mapa a parejas de alumnos. Que usen calculadora y la leyenda del mapa para determinar la distancia de la ruta en línea recta del búho entre el primer y el segundo punto. Después, que determinen las distancias de las rutas por tierra de otros personajes entre el primer y el segundo punto.

   Nota: los alumnos podrían usar pedazos de mondadientes para representar las rutas.

3. Pida a los alumnos que escojan otros pares de puntos en el mapa. Que registren las distancias que recorre el búho entre los dos puntos y las distancias recorridas por los demás personajes de la historia entre los dos puntos.

4. Pida a los alumnos que analicen la información registrada y que escriban acerca de sus observaciones.

How Owl Flies (continued)

Collecting and Organizing Data

While students are measuring and recording their information, ask questions such as:

- What measuring tool are you using? Why? How are you using it? Why is it important to use it in that way?

- What unit of measure are you using? What makes it the most useful for this purpose?

- What information does the legend give you? How are you using that information?

How are you using the calculator to help you find the distances?

How are you using ☐ to represent distances?

What operations are you using to find the distances?

How can you decide if the answer you are getting on the calculator is reasonable or not?

Analyzing Data and Drawing Conclusions

After students have recorded their information, have them work as a whole group to analyze their recording sheets. Ask questions such as:

- How could you describe the way you found the distances on the map?

- Is your group's method the same as other groups' methods?

- How did you use estimation?

- Who had the shortest paths? The longest paths? Why do you think so?

- Do you notice anything interesting in your data? How could you describe it?

How did you use the calculator to help you show the action in your story?

Does the order in which you entered the numbers in your calculator matter to your story? Why or why not?

Continuing the Investigation

Have students select other stories, make up a "map" of the action in the story, make a map legend, and repeat the process of locating points and determining the distance that each character in the story travels between two chosen points.

Cómo vuela el búho (continuación)

Cómo reunir y organizar los datos

Mientras los alumnos miden y registran su información, haga las preguntas siguientes:

- ¿Qué herramienta de medición están usando? ¿Por qué? ¿Por qué es importante usarla así?

- ¿Qué unidad de medida están usando? ¿Por qué es la más útil para este propósito?

- ¿Qué información les da la escala? ¿Cómo están usando esa información?

- ¿Cómo están usando la calculadora para encontrar las distancias?

- ¿Cómo están usando / para representar distancias?

- ¿Qué operaciones están usado para encontrar las distancias?

- ¿Cómo pueden decidir si la respuesta de la calculadora es razonable o no?

Cómo analizar los datos y sacar conclusiones

Después de que los alumnos registren su información, pídales que analicen sus hojas de registro como un solo grupo. Haga las preguntas siguientes:

- ¿Cómo podrían describir la forma en que encontraron las distancias en el mapa?

- ¿El método de su grupo es el mismo que el de otros grupos?

- ¿Cómo usaron la estimación?

- ¿Quiénes tuvieron las rutas más cortas? ¿Las rutas más largas? ¿Por qué piensan eso?

- ¿Observan algo interesante en sus datos? ¿Cómo podrían describirlo?

- ¿Cómo usaron la calculadora para mostrar la acción en su historia?

- ¿El orden en que ingresaron los números en la calculadora tiene alguna importancia en la historia? ¿Por qué?

Cómo continuar la investigación

Que los alumnos seleccionen otras historias, elaboren un "mapa" de la acción en la historia, creen una escala del mapa y repitan el proceso para ubicar puntos y determinar la distancia que recorre cada personaje entre dos puntos seleccionados.

Name:

How Owl Flies
Recording Sheet

Collecting and Organizing Data

The length of Owl's straight-line journey from _____ to _____ is _____ on the map. The actual distance is _____ on the land.

Record the distances of the other characters' journeys in this chart.

Character	From/To	Distance on the Map	Actual Distance on Land
Christopher Robin	From _____ To _____		
Pooh	From _____ To _____		
Eeyore	From _____ To _____		
Tigger	From _____ To _____		
Rabbit	From _____ To _____		

Analyzing Data and Drawing Conclusions

Write about what your table shows you.

Questions we thought of while we were doing this activity:

Nombre:

Cómo vuela el búho
Hoja de registro

Cómo reunir y organizar los datos

La longitud del desplazamiento en línea recta del búho entre _____ y _____ es

de _____ en el mapa. La distancia real es de _____ por tierra.

Registren en el cuadro siguiente las distancias del desplazamiento de los demás personajes.

Personaje	Desde/hasta	Distancia en el Mapa	Distancia real por tierra
Christopher Robin	Desde _____ Hasta _____		
Pooh	Desde _____ Hasta _____		
Eeyore	Desde _____ Hasta _____		
Tigger	Desde _____ Hasta _____		
Rabbit	Desde _____ Hasta _____		

Cómo analizar los datos y sacar conclusiones

Escriban sobre lo que ven en su cuadro.

Preguntas que surgieron mientras realizábamos esta actividad:

How Owl Flies

100 Aker Wood

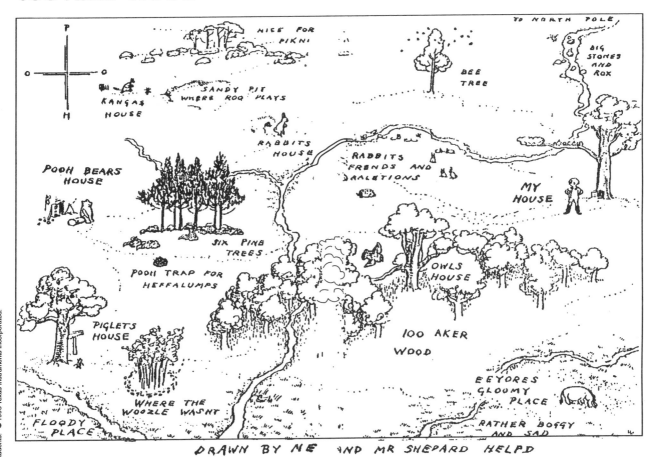

Legend:

0   ½ kilometer

Cómo vuela el búho

100 Aker Wood

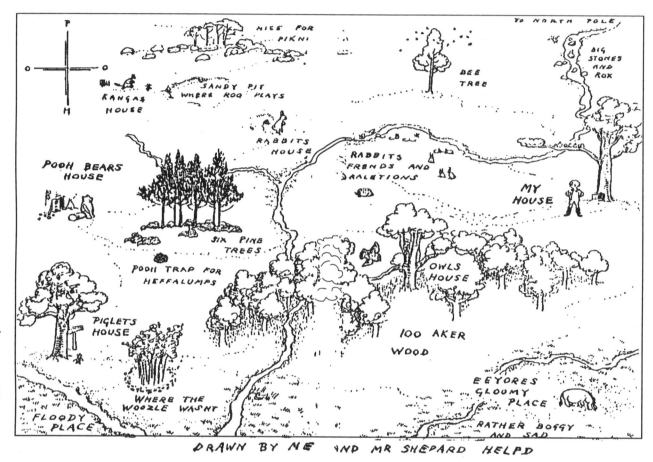

Escala:

0   ½ milla

Extraído de *Winnie the Pooh*, de A. A. Milne, ilustrado por E. H. Shepard. Derechos de autor de E. P. Dutton en 1926 y renovados por A. A. Milne en 1954. Usado con autorización de Dutton Children's Books, división de Penguin Books USA Inc.

Map It!

Math Concepts

- whole numbers
- addition
- linear units of measure
- ratio
- relations and functions

Materials

- TI-108, Math Mate™, Math Explorer™
- **Map It!** recording sheets
- linear measuring tools (rulers, tape measures, string, etc.)
- pencils

Overview

Students will read a story, identify the locations of different places in the story, decide the distance between each set of places in the story, and create a map with a legend.

Introduction

> The **How Owl Flies** activity on page 61 should be completed before beginning this activity.

1. Collect a variety of books that have plots in which characters move from place to place within the story.

   Examples: Some of the Laura Ingalls Wilder's "Little House" books or *The Wizard of Oz* by Frank Baum.

2. With the class, choose a book and then read the story to the class.

   a. Make a transparency, which will be an imaginary map of the action in the story.

   b. Create a legend with a scale for distances.

   c. Identify at least three different locations from the story.

   Examples: In Judith Viorst's *Alexander and the Terrible, Horrible, No Good, Very Bad Day*, locations could include home, Dad's office, the dentist's office, and the shoe store.

   d. Put the places in the story in logical locations on the map.

   e. Use the legend to determine the distances between those places on the map.

3. Have students select a different story and work in pairs to develop a map with a legend to illustrate the primary action in the plot of the new story.

4. Ask students to write about the process they used to make their maps and any mistakes and discoveries they made.

¡Dibuja el mapa!

Conceptos matemáticos

- números enteros
- relación
- suma
- relaciones y funciones
- unidades lineales de medida

Materiales

- TI-108, Math Mate™, Math Explorer™
- hojas de registro de ¡**Dibuja el mapa!**
- herramientas de medición lineal (reglas, cinta métrica, cuerda, etc.)
- lápices

Resumen

Los alumnos leerán una historia, identificarán las ubicaciones de distintos lugares en la historia, decidirán la distancia entre cada grupo de lugares en la historia y crearán un mapa con una escala.

Introducción

Antes de comenzar con esta actividad, se debe completar el ejercicio de **Cómo vuela el búho** en la página 61.

1. Reúna varias historias en las que los personajes se desplacen de un lugar a otro.

    Ejemplos: algunos de los libros de "La pequeña casa en la pradera", de Laura Ingalls Wilder, o *The Wizard of Oz,* de Frank Baum.

2. Con los alumnos, escoja un libro y léales la historia.

    a. Haga una transparencia, que será un mapa imaginario de la acción en la historia.

    b. Cree una escala de distancias.

    c. Identifique al menos tres lugares distintos en la historia.

    Ejemplos: en *Alexander and the Terrible, Horrible, No Good, Very Bad Day,* de Judih Viorst, entre los lugares se podrían incluir la casa, la oficina del papá, la oficina del dentista y la zapatería.

    d. Ponga los lugares de la historia en ubicaciones lógicas dentro del mapa.

    e. Use la escala para determinar las distancias entre los lugares del mapa.

3. Pida a los alumnos que seleccionen una historia diferente y que trabajen en parejas para desarrollar un mapa con una escala para ilustrar la acción primaria en el argumento de la nueva historia.

4. Pida a los alumnos que escriban sobre el proceso que utilizaron para hacer sus mapas y todos los errores y descubrimientos que hicieron.

Map It! (continued)

Collecting and Organizing Data

While students are developing their maps and measuring and recording their information, ask questions such as:

- Which locations did you choose from your story? How did you decide the distances between the locations? How did you determine the relationship between the distances between the places in the story and the distances on your map? How does your legend reflect those relationships?

- What measuring tool are you using? Why? How are you using it? Why is it important to use it in that way?

- What unit of measure are you using? What makes it the most useful for this purpose?

▣ How are you using the calculator to help you find the distances on land and on your map?

▣ What operations are you using on the calculator to help you find the distances?

▣ How can you decide if the answer you are getting on the calculator is reasonable or not?

Analyzing Data and Drawing Conclusions

After students have recorded their information, have them work as a whole group to analyze their recording sheets. Ask questions such as:

- What information did you include in your legend? How did you use that information to create your map?

- Did you choose any distances you were unable to show on your map? Why?

- How could you describe the way you found the distances on the map?

- How did you use estimation?

- Did you have any difficulties making your map? What discoveries did you make?

▣ How did you use the calculator to help you in making the legend for your map?

▣ How does your use of the calculator in this activity compare with the way you used it in **How Owl Flies**?

Continuing the Investigation

Have students select another story, make up a "map" of the action in the story, create a legend for the map, and repeat the process of locating points and determining the distances each character in the story will travel between two points.

¡Dibuja el mapa! (continuación)

Cómo reunir y organizar los datos

Mientras los alumnos desarrollan sus mapas y miden y registran su información, haga las preguntas siguientes:

- ¿Qué puntos escogieron de la historia? ¿Cómo decidieron las distancias entre los puntos? ¿Cómo determinaron la relación entre las distancias entre los lugares de la historia y las del mapa? ¿Cómo refleja su escala esas relaciones?

- ¿Qué herramienta de medición están usando? ¿Por qué? ¿Cómo la están usando? ¿Por qué es importante usarla así?

- ¿Qué unidad de medida están usando? ¿Por qué es la más útil para este propósito?

- ¿Cómo están usando la calculadora para encontrar las distancias por tierra y en su mapa?

- ¿Qué operaciones están usando en la calculadora para encontrar las distancias? ¿Y en el mapa?

- ¿Cómo pueden decidir si la respuesta de la calculadora es razonable o no?

Cómo analizar y sacar conclusiones

Después de que los alumnos hayan registrado su información, pídales que analicen sus hojas de registro como un solo grupo, usando las preguntas siguientes:

- ¿Qué información incluyeron en su escala? ¿Cómo usaron esa información para crear su mapa?

- ¿Escogieron alguna distancia que no pudieran mostrar en su mapa? ¿Por qué?

- ¿Cómo podrían describir la forma en que encontraron las distancias en el mapa?

- ¿Cómo usaron la estimación?

- ¿Tuvieron dificultades para hacer su mapa? ¿Qué descubrimientos hicieron?

- ¿Cómo usaron la calculadora para hacer la escala de su mapa?

- ¿Cómo se compara el uso de la calculadora en esta actividad con la forma en que la usaron en **Cómo vuela el búho**?

Cómo continuar la investigación

Que los alumnos seleccionen otra historia, elaboren un "mapa" de la acción en la historia, creen una escala para el mapa y repitan el proceso de ubicar puntos y determinar las distancias que recorre cada personaje en la historia entre dos puntos.

Name:

Map It!
Recording Sheet

Collecting and Organizing Data

Legend: _____

Locations	Distance on Map	Distance on Land
From _____ To _____		
From _____ To _____		
From _____ To _____		

Use your legend and the information above to draw your map below. Or use the back of this sheet if you need more space.

Analyzing Data and Drawing Conclusions

From this activity, we discovered:

Nombre:

¡Dibuja el mapa!
Hoja de registro

Cómo reunir y organizar los datos

Escala: _____

Ubicaciones	Distancia en el mapa	Distancia por tierra
Desde _____ Hasta _____		
Desde _____ Hasta _____		
Desde _____ Hasta _____		

Use su escala y la información anterior para dibujar el mapa a continuación, o bien utilice el reverso de esta hoja si necesita más espacio.

Cómo analizar los datos y sacar conclusiones

A partir de esta actividad, descubrimos lo siguiente:

Only Half There?

TI-108
MATH MATE
MATH EXPLORER

Math Concepts

- fractions
- ratio
- division
- proportion
- linear measure

Materials

- TI-108, Math Mate™, Math Explorer™
- **Only Half There?** recording sheets
- linear measuring tools (rulers, tape measures, string, etc.)
- pencils or markers
- butcher paper

Overview

Students will use measuring tools and calculators to make half-sized drawings of themselves.

Introduction

1. Read selected portions from *Gulliver's Travels* by A. Benduce, *The Littles* by J. Peterson, or *The Borrowers* by M. Norton. Discuss what the little people might look like.

2. Have students work in pairs. Have one partner lie down on a piece of butcher paper while the other traces around his or her partner's body. Then have the partners trade places.

3. When everyone has a body tracing, have students cut them out and fold them in half from top to bottom. Then give each student another sheet of paper that is half the length of his or her body.

4. Challenge students to draw a half-sized version of themselves. Have students discuss what ideas need to be considered. Have them record their measurements in the table on their recording sheets.

   Examples: What measurements need to be taken? How would those measurements be translated into the half-sized version?

5. Encourage students to use details in their drawings; for example, their facial features, clothing, etc.

Collecting and Organizing Data

While students are working on drawing the half-sized versions of themselves, ask questions such as:

- What measurements are you taking? Why did you choose those?

- What measuring tools are you using? Why did you choose those?

 How are you using the calculator to help you with this problem?

¿Es sólo la mitad?

Conceptos matemáticos

- fracciones
- relación
- división
- proporción
- medida lineal

Materiales

- TI-108, Math Mate™, Math Explorer™
- hojas de registro de **¿Es sólo la mitad?**
- herramientas de medición lineal (reglas, cinta métrica, cuerda, etc.)
- lápices o marcadores de fibra
- papel para envolver

Resumen

Los alumnos utilizarán herramientas de medición y la calculadora para hacer dibujos de sí mismos a la mitad de su tamaño.

Introducción

1. Lea partes seleccionadas de *Gulliver's Travels,* de A. Benduce, *The Littles,* de J. Peterson, o *The Borrowers,* de M. Norton. Analice cómo se vería la gente pequeña de esas historias.

2. Haga que los alumnos trabajen en parejas. Que uno de ellos se acueste sobre un pedazo de papel para envolver mientras el otro dibuja la silueta de su compañero. Después, que cambien de puesto.

3. Cuando todos hayan dibujado su silueta, que los alumnos las recorten y las doblen por la mitad desde arriba hacia abajo. Después, déle a cada alumno otra hoja de papel que sea la mitad de largo que su cuerpo.

4. Pida a los alumnos que hagan un dibujo de sí mismos a la mitad de su tamaño. Que los alumnos analicen las ideas que deben considerar. Que registren sus mediciones en el cuadro de su hoja de registro.

   Ejemplos: ¿qué mediciones deben tomarse? ¿Cómo podrían traducirse esas mediciones en una versión a la mitad del tamaño?

5. Estimule a los alumnos para que utilicen detalles en sus dibujos; por ejemplo, sus características faciales, la ropa, etc.

Cómo reunir y organizar los datos

Mientras los alumnos trabajan en sus dibujos, haga las preguntas siguientes:

- ¿Qué mediciones están tomando? ¿Por qué las escogieron?

- ¿Que herramientas de medición están usando? ¿Por qué las escogieron?

 ¿Cómo están usando la calculadora para ayudarles en este problema?

Only Half There? (continued)

Collecting and Organizing Data (continued)

- How do your actual measurements relate to the measurements in your half-sized picture?

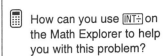 How can you use INT÷ on the Math Explorer to help you with this problem?

- Are there any features that come out looking strange? Why do you think that is happening?

  Note: Many students will try to draw their waist using the measurement of half of the circumference. This makes their waist in the picture appear disproportionately large. Using calipers or shadows projected on a piece of paper can help solve this problem.

Analyzing Data and Drawing Conclusions

After students have drawn their half-sized pictures, have them discuss their results as a whole group. Ask questions such as:

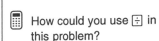 How could you use ÷ in this problem?

- What measuring tools did you use? Why did you choose them?

How could you use ÷ in this problem?

- What patterns, if any, do you see in your pairs of measurements in the table on the recording sheet?

- Why do you think those patterns are occurring?

- Did any parts of your body look distorted? If so, why do you think that happened?

- In a half-sized picture of yourself, is there only half of a full-sized picture? Support your response.

Continuing the Investigation

Have students:

- Draw one-third-sized versions of themselves and discuss their strategies for solving the problem.

- Try to fit drawings of themselves on 5- by 8-inch

¿Es sólo la mitad? (continuación)

Cómo reunir y organizar los datos (continuación)

- ¿Cómo se relacionan las medidas reales con las del dibujo a la mitad del tamaño?

- ¿Alguna de las características resultan extrañas? ¿Por qué creen que sucede esto?

  Nota: muchos alumnos intentarán dibujar su cintura usando la medición de la mitad de la circunferencia. Esto hace que su cintura en el dibujo parezca desproporcionadamente grande. Un compás o sombras proyectadas en un pedazo de papel pueden ayudar a resolver este problema.

🖩 ¿Cómo pueden usar INT÷ en la Math Explorer para ayudarles con este problema?

Cómo analizar los datos y sacar conclusiones

Después de que los alumnos hagan los dibujos a la mitad de su tamaño, pídales que analicen sus resultados como un solo grupo. Haga las preguntas siguientes:

- ¿Qué herramientas de medición utilizaron? ¿Por qué las escogieron?

- ¿Qué patrones, si los hay, observan en sus pares de mediciones en el cuadro de la hoja de registro?

- ¿Por qué piensan que se producen esos patrones?

- ¿Alguna parte de su cuerpo resultó distorsionada? Si fue así, ¿por qué piensan que sucedió?

- En un dibujo de ustedes mismos a la mitad de su tamaño, ¿hay sólo la mitad de un dibujo a tamaño natural? Fundamenten su respuesta.

🖩 ¿Cómo podrían usar ÷ en este problema?

🖩 ¿Cómo podrían usar INT÷ en este problema?

Cómo continuar la investigación

Los alumnos deben:

- Hacer un dibujo de sí mismos a un tercio de su tamaño y analizar sus estrategias para resolver el problema.

- Intentar hacer caber un dibujo de sí mismos en una tarjeta de 5 por 8 pulgadas.

Name:

Only Half There?

Recording Sheet

Collecting and Organizing Data

Body Part	Measurement	$\frac{1}{2}$ Measurement

Analyzing Data and Drawing Conclusions

To make a half-sized drawing of me, I:

Questions we thought of while we were doing this activity:

Nombre:

¿Es sólo la mitad?

Hoja de registro

Cómo reunir y organizar los datos

La parte del cuerpo	Medida	Mitad de la medida

Cómo analizar los datos y sacar conclusiones

Para hacer un dibujo de mí mismo a la mitad de mi tamano, tengo que hacer lo siguiente:

Algunas preguntas que surgieron mientras realizábamos esta actividad:

No More Peas, Please!

Math Concepts

- whole numbers
- ratio
- multiplication
- proportion
- capacity measure

Materials

- TI-108, Math Mate™, Math Explorer™
- **No More Peas, Please!** recording sheets
- measuring tools (string, tape measures, rulers, containers, cubes, dice, marbles, counters, balances, etc.)
- pencils or markers

Overview

Students will use nonstandard units of volume and calculators to estimate the number of peas it would take to fill a room.

Introduction

1. Read the book *Counting on Frank* by Rod Clement to the class. Discuss the different examples of measurement in the story.

   Examples: The length of the ball point pen's line is linear measure; the number of Franks it takes to fill the bedroom is volume.

2. Start two lists on a transparency:
   - Times when units of length are helpful.
   - Times when units of volume are helpful.

3. Ask students to suggest ways to determine the number of peas the artist drew in the illustration.

4. Have students use the number of peas in the illustration to predict the number of peas it would take to fill their classroom.

5. Challenge students to develop a method to find out the number of peas it would take to fill the classroom.

6. Ask students to write a detailed plan for finding the number of peas it would take to fill the classroom.

Collecting and Organizing Data

While students are working on their project, ask questions such as:

- What measurements are you taking? Why did you choose those?

 How are you using the calculator to help solve this problem?

¡Por favor, basta de arvejas!

Conceptos matemáticos

- números enteros
- multiplicación
- medida de capacidad
- relación
- proporción

Materiales

- TI-108, Math Mate™, Math Explorer™
- hojas de registro de **¡Por favor, basta de arvejas!**
- herramientas de medición (cuerda, cinta métrica, reglas, envases, cubos, dados, canicas, contadores, balanzas, etc.)
- lápices o marcadores de fibra

Resumen

Los alumnos utilizarán unidades de volumen no comunes y la calculadora para estimar la cantidad de arvejas que necesitarían para llenar una sala.

Introducción

1. Lea a los alumnos el libro *Counting on Frank*, de Rod Clement. Analice los diferentes ejemplos de medición en la historia.

    Ejemplos: el longitud de la línea de un bolígrafo es una medida lineal; la cantidad de Frank que se necesita para llenar el dormitorio es una medida de volumen.

2. Haga dos listas en una transparencia:

    - Las veces en que se necesitan unidades de longitud.
    - Las veces en que se necesitan unidades de volumen.

3. Pida a los alumnos que sugieran formas para determinar el número de arvejas que dibujó el artista en la ilustración.

4. Que los alumnos usen la cantidad de arvejas en la ilustración para predecir el número de arvejas que se necesitarían para llenar la sala de clases.

5. Que los alumnos desarrollen un método para determinar el número de arvejas que se necesitarían para llenar la sala de clases.

6. Pida a los alumnos que escriban un plan detallado para encontrar el número de arvejas que se necesitarían para llenar la sala de clases.

Cómo reunir y organizar los datos

Mientras los alumnos trabajan en su proyecto, haga las preguntas siguientes:

- ¿Qué mediciones están tomando? ¿Por qué las escogieron?

 ¿Cómo están usando la calculadora para resolver este problema?

No More Peas, Please! (continued)

Collecting and Organizing Data (continued)

- What measuring tools are you using? Why did you choose those?

- Are there any measurement tools on the supply table you don't think would be helpful in solving this problem? Why?

How will you decide if the answer you come up with is reasonable?

Analyzing Data and Drawing Conclusions

After students have described the method they would use to find the number of peas it would take to fill the room, have them discuss their results as a whole group. Ask questions such as:

- Was the illustration in the book helpful in designing a way of finding the number of peas it will take to fill this room? Why or why not?

- What measuring tools did you use? Why did you choose them?

- What measuring tools would not be helpful? Why?

- What do all of the helpful measuring tools seem to have in common?

- What was the most difficult part of this problem? Why?

- If we were to choose one method of solving this problem from among all of those suggested, which should it be? Be prepared to support your suggestion with sound, logical reasons.

What problems did you experience using the calculator in this problem? How did you solve those problems?

Continuing the Investigation

Have students:

- Test the plan they designed for finding the number of peas.

- Make up other measurement problems using the story *Counting on Frank*. Have students write each problem on one side of a card and write one method of solving it on the reverse side.

¡Por favor, basta de arvejas! (continuación)

Cómo reunir y organizar los datos (continuación)

- ¿Qué herramientas de medición están usando? ¿Por qué las escogieron?

- ¿Hay alguna herramienta de medición en el cuadro de materiales que, según ustedes, no serían útiles para resolver este problema? ¿Por qué?

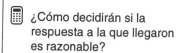 ¿Cómo decidirán si la respuesta a la que llegaron es razonable?

Cómo analizar los datos y sacar conclusiones

Después de que los alumnos describan el método que utilizarían para encontrar el número de arvejas que se necesitarían para llenar la sala, pídales que analicen sus resultados como un solo grupo. Haga las preguntas siguientes:

- ¿La ilustración del libro sirvió para diseñar una forma para definir el número de arvejas que se necesitarían para llenar la sala? ¿Por qué?

- ¿Qué herramientas de medición utilizaron? ¿Por qué las escogieron?

- ¿Qué herramientas de medición no fueron útiles? ¿Por qué?

- ¿Qué parecen tener en común todas las herramientas de medición útiles?

- ¿Cuál fue la parte más difícil de este problema? ¿Por qué?

- Si tuviéramos que escoger un método para resolver este problema de entre todos los sugeridos, ¿cuál sería? Estén preparados para fundamentar su respuesta con razones claras y lógicas.

¿Qué dificultades tuvieron con la calculadora en este problema? ¿Cómo las solucionaron?

Cómo continuar la investigación

Los alumnos deben:

- Probar el plan que diseñaron para determinar el número de arvejas.

- Resolver otros problemas de medición usando la historia *Counting on Frank*. Que los alumnos escriban cada problema en un lado de la tarjeta y un método para resolverlo en el reverso de la misma.

Name:

No More Peas, Please!
Recording Sheet

Collecting and Organizing Data:

Materials we will need:

To find out how many peas it will take to fill this room, we will:

Analyzing Data and Drawing Conclusions

We think the method we chose will work because:

Questions we thought of while we were doing this activity:

Nombre:

¡Por favor, basta de arvejas!
Hoja de registro

Cómo reunir y organizar los datos

Materiales que necesitamos:

Para determinar la cantidad de arvejas que se necesitan para llenar esta sala, tendremos que hacer lo siguiente:

Cómo analizar los datos y sacar conclusiones

Pensamos que el método que escogimos funcionará porque:

Preguntas que surgieron mientras realizábamos esta actividad:

Do Centimeters Make Me Taller?

Math Concepts

- fractions
- decimals
- division
- linear measure
- ratio
- proportion

Materials

- Math Explorer™
- **Do Centimeters Make Me Taller?** recording sheets
- linear measuring tools (rulers, tape measures, string, etc.)
- pencils and large paper

Overview

Students will use measuring tools and calculators to find pairs of measurements and form ratios. Then they will compare the ratios and make conclusions about using different units to measure lengths.

Introduction

1. Have each student lie down on the floor or on a piece of paper and mark the length of his or her height.

2. Then have each student mark the length of his or her foot, use the length of the foot to measure his or her height, and record the measurement on the chart provided on the recording sheet.

3. Next, have each student measure his or her height and the length of his or her foot using a standard unit of measure (inches or centimeters) and record the results on the chart.

4. Discuss the idea of ratio as a number that describes the comparison of two quantities. This number can be expressed as a fraction or a decimal.

   Example: If your foot is 20 cm and your height is 120 cm, the ratio of your foot to your height is 20 cm/120 cm or 1/6.

5. Have students use the $\div$ key or the F⊃D key to change their foot/height ratios from fraction form to decimal form. Have students compare their two ratios and discuss how they are alike and how they are different.

6. Have students select another nonstandard unit of measure (hand, pencil, eraser, etc.), use it to measure both their foot and height, and record their measurements on the chart.

7. Have each student measure his or her foot and height with some other standard unit of measure and record the measurements on the chart.

¿Los centímetros me hacen más alto?

Conceptos matemáticos

- fracciones
- decimales
- división
- medida lineal
- relación
- proporción

Materiales

- Math Explorer™
- hojas de registro de **¿Los centímetros me hacen más alto?**
- herramientas de medición lineal (reglas, cinta métrica, cuerda, etc.)
- lápices y un papel grande

Resumen

Los alumnos utilizarán herramientas de medición y la calculadora para encontrar pares de medidas y relaciones de forma. Después, compararán las relaciones y sacarán conclusiones acerca del uso de diferentes unidades para medir longitudes.

Introducción

1. Que cada alumno se acueste en el piso o sobre un pedazo de papel y marque la longitud de la altura de su cuerpo.

2. Después, que cada alumno marque la longitud de su pie, use ese largo para medir su altura y registre la medición en el cuadro de la hoja de registro.

3. En seguida, que cada alumno mida su altura y la longitud de su pie usando una unidad de medida normal (centímetros o pulgadas) y que registre los resultados en el cuadro.

4. Analice la idea de relación como un número que describe la comparación de dos cantidades. Este número se puede expresar como una fracción o como un decimal.

   Ejemplo: si su pie tiene 20 cm y su altura es de 120 cm, la relación de su pie con respecto a su altura es de 20 cm/120 cm o 1/6.

5. Que los alumnos utilicen las teclas ⊡ o ⌊F⊃D⌋ para cambiar sus relaciones pie/altura desde una forma de fracción a una de decimal. Que los alumnos comparen sus dos relaciones y analicen en qué se parecen y difieren.

6. Que los alumnos seleccionen otra unidad no común de medida (mano, lápiz, borrador, etc.), la usen para medir su pie y su altura, y registren sus medidas en el cuadro.

7. Que cada alumno mida su pie y altura con una unidad de medida normal y registre las mediciones en el cuadro.

Do Centimeters Make Me Taller? (continued)

Introduction (continued)

8. Have students find the ratios of these pairs of measurements and compare their values.

9. Have students repeat this process with one more nonstandard unit and one more standard unit of measure and then discuss their results.

Example:

Unit of Measure	1st Measure	2nd Measure	Ratio of 1st to 2nd	Ratio in Decimal Form
Nonstandard Unit: *my foot*	*My foot* is: *1 foot* (Nonstandard Units)	*My height* is: *6 feet* (Nonstandard Units)	$\frac{1\ foot}{6\ feet}$	0.1666667
Standard Unit: *cm*	*My foot* is: *25 cm* (Standard Units)	*My height* is: *140 cm* (Standard Units)	$\frac{25\ cm}{140\ cm}$	0.1785714
Nonstandard Unit: *pencil*	*My foot* is: *2 pencils* (Nonstandard Units)	*My height* is: *13 pencils* (Nonstandard Units)	$\frac{2\ pencils}{13\ pencil}$	0.1538462
Standard Unit: *in*	*My foot* is: *10 in* (Standard Units)	*My height* is: *64 in* (Standard Units)	$\frac{10\ in}{64\ in}$	0.15625

Collecting and Organizing Data

While students take measurements to generate the chosen ratios, ask questions such as:

*   What kinds of units are you using to make your measurements?

*   How do your ratios of measurements with nonstandard units compare to your ratios with standard units?

*   How close do you want your ratios to be to accept them as the "same" ratio? What does this mean about the units you are using?

How are you using the calculator to help you with this problem?

How can you use ÷ on the Math Explorer to help you look at ratios?

How can you use F⊂D on the Math Explorer to help you look at ratios?

How can you use Fix to help you decide if your ratios are "close"?

¿Los centímetros me hacen más alto?

(continuación)

Introducción (continuación)

8. Que los alumnos determinen las relaciones de estos pares de mediciones y comparen sus valores.

9. Que los alumnos repitan este proceso con otra unidad de medida no común y otra común y que después analicen sus resultados.

Ejemplo:

Unidad de medida	1ª medida	2ª medida	Relacion entre la 1ª y la 2ª medida	Relacion en forma decimal
Unidad no común: *mi pie*	*Mi pie* es: *1 pie* (unidades no comunes)	*Mi altura* es: *6 pies* (unidades no comunes)	*1 pie* *6 pies*	*0.1666667*
Unidad común: *cm*	*Mi pie* es: *25 cm* (unidades comunes)	*Mi altura* es: *140 cm* (unidades comunes)	*25 cm* *140 cm*	*0.1785714*
Unidad no común: *lápiz*	*Mi pie* es: *2 lápices* (unidades no comunes)	*Mi altura* es: *13 lápices* (unidades no comunes)	*2 lápices* *13 lápices*	*0.1538462*
Unidades comúnes: *pulgada*	*Mi pie* es: *10 pulgadas* (unidades comunes)	*Mi altura* es: *64 pulgadas* (unidades comunes)	*10 pulgadas* *64 pulgadas*	*0.15625*

Cómo reunir y organizar los datos

Mientras los alumnos toman mediciones para generar las relaciones escogidas, haga las preguntas siguientes:

- ¿Qué tipos de unidades están usando para tomar las medidas?

- ¿Cómo se comparan sus relaciones de medidas con unidades no comunes y las relaciones con unidades comunes?

- ¿Qué aproximación deben tener sus relaciones para que las consideren "iguales"? ¿Cómo se traduce esto con respecto a las unidades que están usando?

🔲 ¿Cómo están usando la calculadora con este problema?

🔲 ¿Cómo pueden usar ⊟ en Math Explorer para examinar las relaciones?

🔲 ¿Cómo pueden usar F⊂D en el Math Explorer para examinar las relaciones?

🔲 ¿Cómo pueden usar Fix para decidir si sus relaciones son "aproximadas"?

Do Centimeters Make Me Taller? (continued)

Analyzing Data and Drawing Conclusions

After students have found several pairs of measurements, have them discuss their results as a whole group. Ask questions such as:

- What measuring tools did you use? Why did you choose them?

- What patterns, if any, do you see in your pairs of measurements?

- Why do you think those patterns are occurring?

- How close did you decide the ratios had to be in order to be the "same"?

- Why was it necessary to allow the ratios to be "close" rather than exactly equal?

- If someone else made your same pairs of measurements, would their data come out exactly the same as yours? Why or why not? What would be different, if anything?

- Does measuring your height in centimeters make you taller than measuring your height in inches? Why or why not?

- How did you use $\boxed{\text{Fix}}$ in this problem?

- How could you use $\boxed{/}$ in this problem?

- How could you use $\boxed{\text{F}\supset\text{D}}$ in this problem?

- How could you use $\boxed{\div}$ in this problem?

- Would you want to use $\boxed{\text{INT}\div}$ in this problem? Why or why not?

Continuing the Investigation

Have students:

- Pick a ratio and try to find pairs of measurements that will form ratios close to the one they picked.

- Trade recording sheets, redo the pairs of measurements on the recording sheet they received, and see how closely their results match the first group's results.

- Use nonstandard and standard units to measure pairs of objects other than their feet and heights to see if the same pattern occurs in the ratios.

¿Los centímetros me hacen más alto?

(continuación)

Cómo analizar los datos y sacar conclusiones

Después de que los alumnos encuentren varios pares de medidas, pídales que analicen sus resultados como un solo grupo. Haga preguntas siguientes:

- ¿Qué herramientas de medición utilizaron? ¿Por qué las escogieron?

- ¿Qué patrones, si los hay, observan en sus pares de medidas?

- ¿Por qué piensan que se producen estos patrones?

- ¿Cúan aproximadas decidieron que debían ser sus relaciones para considerarlas "iguales"?

- ¿Por qué fue necesario permitir que las relaciones fueran "aproximadas" en vez de exactamente iguales?

- Si otra persona hizo sus mismos pares de medidas, ¿los datos de esa persona serían exactamente los mismos que los suyos? ¿Por qué? ¿Qué elementos serían diferentes, si los hubiera?

- ¿Medir su altura en centímetros los hace más altos que al hacerlo en pulgadas? ¿Por qué?

🔲 ¿Cómo utilizaron Fix en este problema?

🔲 ¿Cómo utilizaron / en este problema?

🔲 ¿Cómo utilizaron F⊂D en este problema?

🔲 ¿Cómo utilizaron ÷ en este problema?

🔲 ¿Usarían INT÷ en este problema? ¿Por qué?

Cómo continuar la investigación

Los alumnos deben:

- Seleccionar una relación e intentar encontrar pares de medidas que formen relaciones aproximadas a la que seleccionaron.

- Intercambiar hojas de registros, rehacer los pares de medidas en la hoja de registro que recibieron y ver con qué aproximación sus resultados concuerdan con los del primer grupo.

- Usar unidades de medida comunes y no comunes para medir pares de objetos que no sean su altura o su pie, para ver si se produce el mismo patrón en las relaciones.

Name:

Do Centimeters Make Me Taller?
Recording Sheet

Collecting and Organizing Data

Unit of Measure	1st Measure	2nd Measure	Ratio of 1st to 2nd	Ratio in Decimal Form
Nonstandard Unit:_____	_____ is: _____(Nonstandard Units)	_____ is: _____(Nonstandard Units)		
Standard Unit:_____	_____ is: _____(Standard Units)	_____ is: _____(Standard Units)		
Nonstandard Unit:_____	_____ is: _____(Nonstandard Units)	_____ is: _____(Nonstandard Units)		
Standard Unit:_____	_____ is: _____(Standard Units)	_____ is: _____(Standard Units)		
Nonstandard Unit:_____	_____ is: _____(Nonstandard Units)	_____ is: _____(Nonstandard Units)		
Standard Unit:_____	_____ is: _____(Standard Units)	_____ is: _____(Standard Units)		

Analyzing Data and Drawing Conclusions

What we noticed about our pairs of ratios:

What we think that means:

Questions we thought of while we were doing this activity:

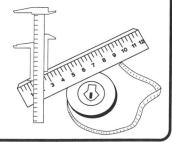

Nombre:

¿Los centímetros me hacen más alto?

Hoja de registro

Cómo reunir y organizar los datos

Unidad de medida	1ª medida	2ª medida	Relación entre 1ª y 2ª medida	Relación en forma decimal
Unidad no común: _____	_____ es: _____(Unidades no común:)	_____ es: _____(Unidades no común:)		
Unidad común:_____	_____ es: _____(Unidades común:)	_____ es: _____(Unidades común:)		
Unidad no común: _____	_____ es: _____(Unidades no común:)	_____ es: _____(Unidades no común:)		
Unidad común_____	_____ es: _____(Unidades común:)	_____ es: _____(Unidades común:)		
Unidad no común: _____	_____ es: _____(Unidades no común:)	_____ es: _____(Unidades no común:)		
Unidad común:_____	_____ es: _____(Unidades común:)	_____ es: _____(Unidades común:)		

Cómo analizar los datos y sacar conclusiones

Observamos lo siguiente acerca de nuestros pares de relaciones

Lo que pensamos significa lo siguiente:

Preguntas que surgieron mientras realizábamos esta actividad:

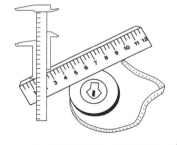

What's My Ratio?

Math Concepts
- fractions
- decimals
- division
- linear measure
- ratio
- proportion
- similarity

Materials
- Math Explorer™
- **What's My Ratio?** recording sheets
- centimeter grid paper
- rulers or other linear measuring tools
- pencils
- pictures with enlarged or reduced copies

Overview

Students will use linear measurement and calculators to investigate proportionality and determine the constant ratio between similar figures.

Introduction

1. Read appropriate sections of *Goosebumps — Monster Blood III* (Chapters 15 and 16) by R.L. Stine or *The Shrinking of Treehorn* by F. Heide. Have students discuss what would happen to a picture in the pocket of someone who is shrunk or "blown up."

2. Have students draw a simple picture on centimeter grid paper (or use the centimeter grid provided on page 83). Have them decide to either increase or decrease the size of the picture, predict what they think the dimensions will be in the increased or decreased version, and discuss their reasoning.

3. Have students draw the increased or decreased picture on grid paper, keeping the picture's original shape, to test their predictions.

4. Discuss the use of ratio (comparing the measurements of corresponding parts) to describe how the picture has been increased or decreased.

   Example: If a line in the first picture is 3 cm and the corresponding line in the second picture is 6 cm, the ratio of the first picture to the second picture is 3 to 6 or three-sixths (or one-half, in simplest form).

5. Divide students into groups. Give each group an interesting picture (or use those provided on page 82) and a reduced or enlarged copy of the same picture. Ask each group to measure several pairs of corresponding parts on the two pictures, record their data on the recording sheet, and make some conclusions about copies of pictures.

¿Cuál es mi relación?

Conceptos matemáticos

- fracciones
- decimales
- división
- medida lineal
- relación
- proporción
- similitud

Materiales

- Math Explorer™
- hojas de registro de **¿Cuál es mi relación?**
- papel cuadriculado con cuadrados de un centímetro
- reglas u otras herramientas de medición lineal
- lápices
- imágenes con copias ampliadas o reducidas

Resumen

Los alumnos utilizarán la medición lineal y la calculadora para investigar la proporcionalidad y determinar la relación constante entre figuras similares.

Introducción

1. Lea algunas partes adecuadas de *Goosebumps — Monster Blood III*, de R.L. Stine o de *The Shrinking of Treehorn,* de F. Heide. Que los alumnos analicen lo que sucedería a una imagen en el bolsillo de una persona que fuera reducida o agrandada.

2. Que los alumnos hagan un dibujo simple en papel cuadriculado con cuadrados un centímetro (o que usen la cuadrícula de la página 83). Pídales que decidan si aumentar o disminuir el tamaño del dibujo, que predigan cuáles serán las dimensiones de la versión ampliada o reducida, y que analicen su razonamiento.

3. Que los alumnos dibujen la imagen ampliada o reducida en papel cuadriculado, manteniendo la forma original de la misma para probar sus predicciones.

4. Analice el uso de la relación (comparando las medidas de las partes correspondientes) para describir cómo la imagen se ha ampliado o reducido.

   Ejemplo: si una línea de la primera imagen es de 3 cm y la línea correspondiente en la segunda imagen es de 6 cm, la relación entre la primera y la segunda imagen es de 3 a 6 o tres sextos (o una mitad, en forma más simple).

5. Divida a los alumnos en grupos. Entregue a cada grupo una imagen interesante (o use las de la página 82) y una copia reducida o ampliada de la misma. Que cada grupo mida varios pares de partes que correspondan en dos imágenes, registre sus datos en la hoja de registro y saque algunas conclusiones acerca de las copias de las imágenes.

What's My Ratio? (continued)

Collecting and Organizing Data

While students take measurements to generate data for comparing the ratios, ask questions such as:

- How are you going to compare these two pictures?

- What is your estimate of the change in size?

- Does that estimate make sense? Why or why not?

- How would you express the change as a comparison between the two pictures?

- What kind of attributes could you compare?

- Is it important to compare the same things in the two pictures? Why or why not?

- What have you done previously in mathematics that might apply to this problem?

- How will you explain your strategy to the rest of the class?

- Would your strategy work for any picture? If so, why? If not, why not?

- What patterns, if any, do you see in the data?

- What conjectures have you made from the patterns in the data?

How are you using the calculator to help you with this problem?

How can you use FCD on the Math Explorer to help you look for patterns?

How can you use the calculator to compare fraction and decimal representations of ratios?

Analyzing Data and Drawing Conclusions

After students have taken several measurements and compared several ratios in their pictures, have them discuss their results as a whole group. Ask questions such as:

- Did your results match your estimates? Why or why not?

- How did you determine the ratio between the two figures?

- How did you use measuring tools to help find the ratios?

- What problems did you encounter, and how did you solve them?

- What mathematics did you use to find the ratios?

How could you use / in this problem?

How could you use FCD in this problem?

How could you use ÷ in this problem?

Would you want to use INT÷ in this problem? Why or why not?

¿Cuál es mi relación? (continuación)

Cómo reunir y organizar los datos

Mientras los alumnos toman medidas para generar datos y así comparar las relaciones, haga las preguntas siguientes:

- ¿Cómo van a comparar estas dos imágenes?

- ¿Cuál es su estimación del cambio de tamaño?

- ¿Tiene algún sentido esta estimación? ¿Por qué?

- ¿Cómo expresarían el cambio como una comparación entre las dos imágenes?

- ¿Qué tipos de atributo podrían comparar?

- ¿Es importante comparar las mismas cosas en las dos imágenes? ¿Por qué?

- ¿Qué han hecho anteriormente en matemáticas que pudieran aplicar a este problema?

- ¿Cómo explicarán su estrategia al resto de sus compañeros?

- ¿Su estrategia funcionaría para cualquier imagen? Si es así, ¿por qué? Si no ocurre así, ¿por qué?

- ¿Qué patrones, si los hay, observan en los datos?

- ¿Qué conjeturas han elaborado a partir de los patrones en los datos?

⊞ ¿Cómo están usando la calculadora para resolver este problema?

⊞ ¿Cómo pueden utilizar F⊃D en la Math Explorer para buscar patrones?

⊞ ¿Cómo pueden emplear la calculadora para comparar representaciones de relaciones en forma de fracción o de decimal?

Cómo analizar los datos y sacar conclusiones

Después de que los alumnos tomen varias mediciones y comparen varias relaciones en sus imágenes, pídales que analicen sus resultados. Haga las preguntas siguientes:

- ¿Sus resultados concordaron con sus estimaciones? ¿Por qué?

- ¿Cómo determinaron la relación entre las dos figuras?

- ¿Cómo usaron las herramientas de medición para encontrar las relaciones?

- ¿Qué problemas encontraron y cómo los resolvieron?

- ¿Qué procedimientos matemáticos utilizaron para encontrar las relaciones?

⊞ ¿Cómo podrían usar / en este problema?

⊞ ¿Cómo podrían usar F⊃D en este problema?

⊞ ¿Cómo podrían usar ÷ en este problema?

⊞ ¿Cómo usarían INT÷ en este problema? ¿Por qué?

What's My Ratio? (continued)

Analyzing Data and Drawing Conclusion (continued)

- What patterns did you find in the ratios?

- Why do you think those patterns exist?

- What do you think would happen if you changed the values of any of the ratios between corresponding parts in a pair of pictures? Why do you think that would happen?

Continuing the Investigation

Have students:

- Create their own drawings, trade drawings with other students, and increase or decrease the drawings by a given ratio.

- Investigate the ratio between the areas of the two pictures. Is it the same as the ratio between the linear dimensions? Why or why not?

  Note: Investigate with simple squares to form a conjecture.

¿Cuál es mi relación? (continuación)

Cómo analizar los datos y sacar conclusiones (continuación)

- ¿Qué patrones encontraron en las relaciones?

- ¿Por qué piensan que existen esos patrones?

- ¿Qué piensan que sucedería si cambiaran los valores de cualquiera de las relaciones entre las partes correspondientes en un par de imágenes? ¿Por qué creen se sucedería esto?

Cómo continuar la investigación

Los alumnos deben:

- Crear sus propios dibujos, intercambiarlos con los demás alumnos y ampliar o reducir los dibujos según una relación determinada?

- Investigar la relación entre las superficies de las dos imágenes. ¿Es la misma que la relación entre las dimensiones lineales? ¿Por qué?

  Nota: investigue con cuadrados simples para elaborar una conjetura.

Name:

What's My Ratio?
Recording Sheet

Collecting and Organizing Data

We measured a picture of a :_____

Part of the Picture that We Measured	Measurement in 1st Picture	Measurement in 2nd Picture	Ratio in Fraction Form	Ratio in Decimal Form

Analyzing Data and Drawing Conclusions

Questions we thought of while we were doing this activity:

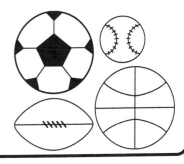

Nombre:

¿Cuál es mi relación?

Hoja de registro

Cómo reunir y organizar los datos

Medí una imagen de: _____

Parte de la imagen que medí	Medida de la 1ª imagen	Medida de la 2ª imagen	Relación en forma de fracción	Relación en forma decimal

Cómo analizar los datos y sacar conclusiones

Algunas preguntas que surgieron mientras realizábamos esta actividad:

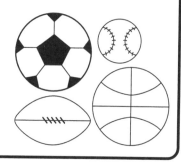

What's My Ratio?

Sample Pictures to Shrink or Enlarge

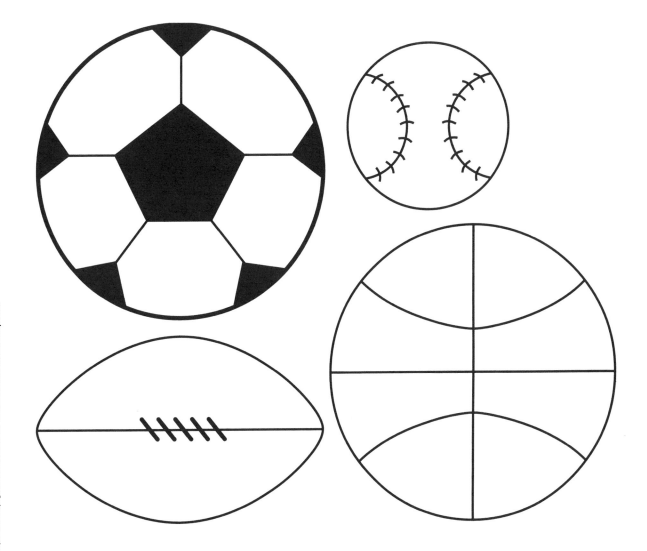

¿Cuál es mi relación?

Modelos de imágenes para ampliar o reducir

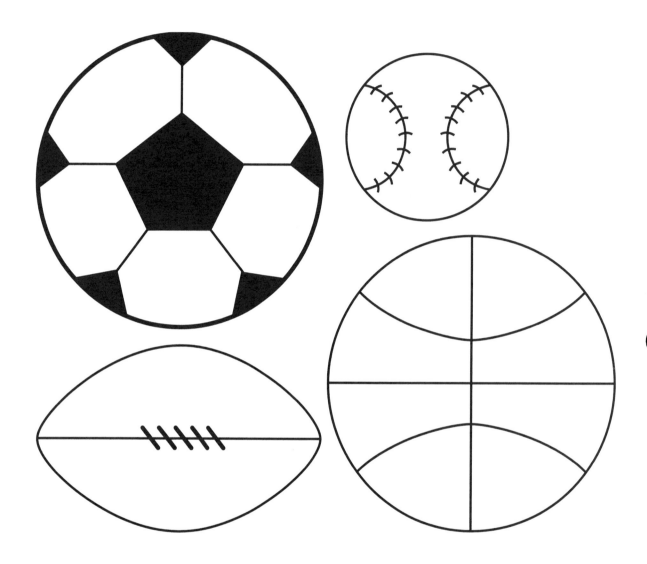

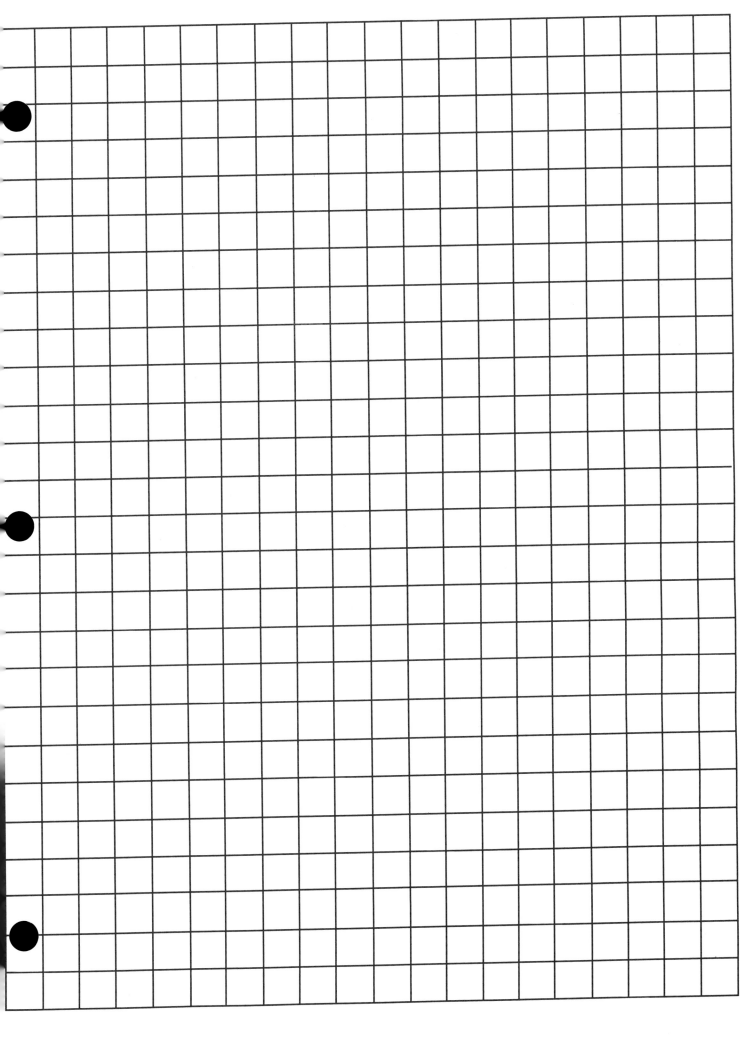

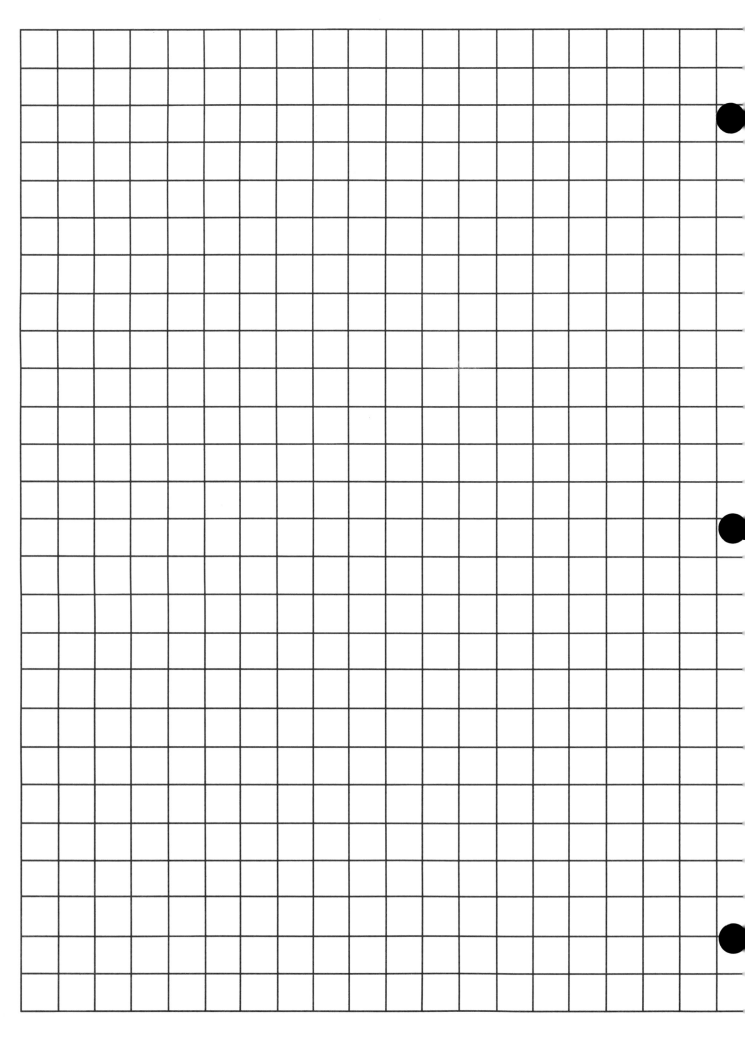

Ratios in Regular Polygons

Math Concepts

- fractions
- decimals
- linear measure
- ratio
- proportion
- segments
- polygons
- congruence
- similarity

Materials

- Math Explorer™
- **Ratios in Regular Polygons** recording sheets
- rulers and pencils

Overview

Students will use linear measurement and calculators to investigate the ratios between corresponding parts of regular polygons.

Introduction

1. Have students draw several triangles, compare their triangles with those of other students, and look for any similarities among all the triangles.

   Note: There should be very few similarities.

2. Next, have students draw several equilateral triangles, compare their triangles with those of other students and look for similarities.

   Note: They are all the same shape but different sizes.

3. Have students do the same experiment with rectangles, and then squares.

   Note: The rectangles come in all shapes; the squares are all the same shape but different sizes.

4. Introduce the term *similar figures* to mean "having the same shape but not necessarily the same size."

5. Give students the picture of several different-sized squares (see page 87). Have students measure the length of the diagonal and the perimeter of each square, record their findings on the recording sheet, and look for patterns.

6. Have students record the same data for other regular polygons of several different sizes and look for patterns. Regular hexagons, pentagons, and octagons are on page 87.

Relaciones en polígonos regulares

Conceptos matemáticos

- fracciones
- decimales
- medida lineal
- relación
- proporción
- segmentos
- polígonos
- congruencia
- similitud

Materiales

- Math Explorer™
- hojas de registro de **Relaciones en polígonos regulares**
- reglas y lápices

Resumen

Los alumnos usarán la medición lineal y la calculadora para investigar las relaciones entre las partes correspondientes de polígonos regulares.

Introducción

1. Haga que los alumnos dibujen varios triángulos, comparen sus triángulos con los de los demás alumnos y busquen similitudes entre todos los triángulos.

   Nota: debería haber muy pocas similitudes.

2. Después, que los alumnos dibujen varios triángulos equilateros, comparen sus triángulos con los de los demás estudiantes y busquen similitudes.

   Nota: todos tienen la misma forma, pero distinto tamaño.

3. Que los estudiantes hagan el mismo experimento con rectángulos y después cuadrados.

   Nota: los rectángulos vienen en todas las formas; los cuadrados tienen todos la misma forma, pero distinto tamaño.

4. Introduzca el término *figuras similares* para referirse a figuras "que tienen la misma forma, pero no necesariamente el mismo tamaño".

5. Entregue a los alumnos la imagen de varios cuadrados (vea la página 87). Que midan la longitud de la diagonal y el perímetro de cada cuadrado, anoten sus descubrimientos en la hoja de registro y busquen patrones.

6. Que los alumnos anoten los mismos datos para otros polígonos regulares de varios tamaños distintos y que busquen patrones. Los octágonos, pentágonos y hexágonos regulares están en la página 87.

Ratios in Regular Polygons (continued)

Collecting and Organizing Data

While students generate data for the different sets of similar figures, ask questions such as:

- How are all of these squares (or hexagons, pentagons, etc.) alike?

- How are you measuring the diagonals?

- How are you measuring the perimeters?

- How do you know your measurements are reasonable?

- Does it matter if you measure in inches or centimeters? Why or why not?

- What patterns do you see? Why do you think those patterns are occurring?

▦ How can you use division with the calculator to help you look for patterns?

▦ How can you use FCD on the Math Explorer to help you look for patterns?

▦ How can you judge if what you see on your calculator is reasonable?

▦ How can you use the calculator and the patterns you see to help you predict measurements?

Analyzing Data and Drawing Conclusions

After students have made and compared several sets of measurements, have them discuss their results as a whole group. Ask questions such as:

- Did your data turn out exactly like everyone else's? Why or why not?

- What patterns do you see in your data?

- How are the diagonals and the perimeters of squares related to each other? Of regular pentagons? Of regular hexagons? Of regular octagons?

- From the patterns in your data, what conjectures can you make about measurements in similar figures?

▦ What operations or keys did you use on the calculator to help you find patterns in this activity? Why did you choose those operations or keys?

▦ How did you determine if your calculator results were reasonable?

Continuing the Investigation

Have students:

- Look for relationships between measurements of other parts of similar figures; for example, perimeter and area.

- Investigate similar figures other than regular polygons; for example, nonsquare rectangles that are the same shape, scalene triangles that are the same shape, etc.

Relaciones en polígonos regulares

(continuación)

Cómo reunir y organizar los datos

Mientras los alumnos generan datos para los distintos grupos de figuras similares, haga las preguntas siguientes:

- ¿En qué se parecen todos estos cuadrados (o hexágonos, pentágonos, etc.)?

- ¿Cómo están midiendo las diagonales?

- ¿Cómo está midiendo los perímetros?

- ¿Cómo saben si sus medidas son razonables?

- ¿Tiene alguna importancia si miden en pulgadas o centímetros? ¿Por qué?

- ¿Qué patrones observan? ¿Por qué piensan que se producen estos patrones?

▢ ¿Cómo pueden usar la división con la calculadora para buscar patrones?

▢ ¿Cómo pueden usar [F⊂D] en Math Explorer para buscar patrones?

▢ ¿Cómo pueden juzgar si lo que observan en la calculadora es razonable?

▢ ¿Cómo pueden usar la calculadora y los patrones que observan para predecir medidas?

Cómo analizar los datos y sacar conclusiones

Después de que los alumnos hagan y comparen varios grupos de medidas, pídales que analicen sus resultados. Haga las preguntas siguientes:

- ¿Sus datos resultaron exactamente iguales a los de los demás? ¿Por qué?

- ¿Qué patrones observan en sus datos?

- ¿Cómo se relacionan las diagonales y los perímetros de los cuadrados? ¿Los pentágonos regulares? ¿Los hexágonos regulares? ¿Los octágonos regulares?

- A partir de los patrones de sus datos, ¿qué conjeturas pueden formular acerca de las medidas en figuras similares?

▢ ¿Qué operaciones o teclas usaron en la calculadora para encontrar patrones en esta actividad? ¿Por qué escogieron estas operaciones o teclas?

▢ ¿Cómo determinaron si los resultados de su calculadora eran razonables?

Cómo continuar la investigación

Los alumnos deben:

- Buscar relaciones entre las medidas de otras partes de figuras similares; por ejemplo, perímetro y superficie.

- Investigar figuras similares distintas a los polígonos regulares; por ejemplo, rectángulos no cuadrados que tengan la misma forma, triángulos escalenos que tengan la misma forma, etc.

Name:

Ratios in Regular Polygons
Recording Sheet

Collecting and Organizing Data

Polygon investigated:_____

Measurement of Perimeter	Measurement of Diagonal	Ratio of Perimeter to Diagonal	Ratio in Decimal Form

Analyzing Data and Drawing Conclusions

What we noticed about the ratios of the different-sized polygons:

Questions we thought of while we were doing this activity:

Nombre:

Relaciones en polígonos regulares
Hoja de registro

Cómo reunir y organizar los datos

Polígono investigado:_____

Medida del perímetro	Medida de la diagonal	Relación entre el perímetro y la diagonal	Relación en forma decimal

Cómo analizar los datos y sacar conclusiones

Observamos lo siguiente acerca de las relaciones de los polígonos de diferentes tamaños:

Algunas preguntas que surgieron mientras realizábamos esta actividad:

Ratios in Regular Polygons

Regular Polygons of Different Sizes

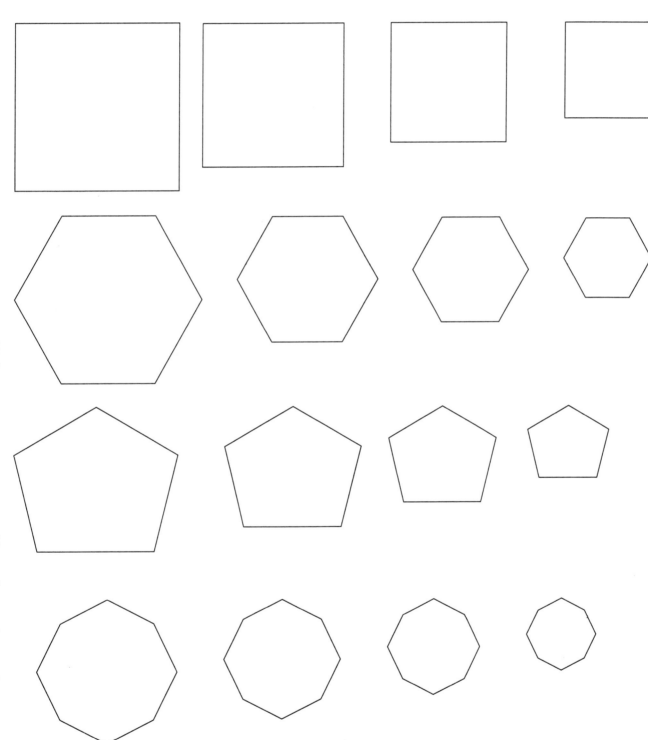

Relaciones en polígonos regulares

Polígonos regulares de tamaños diferentes:

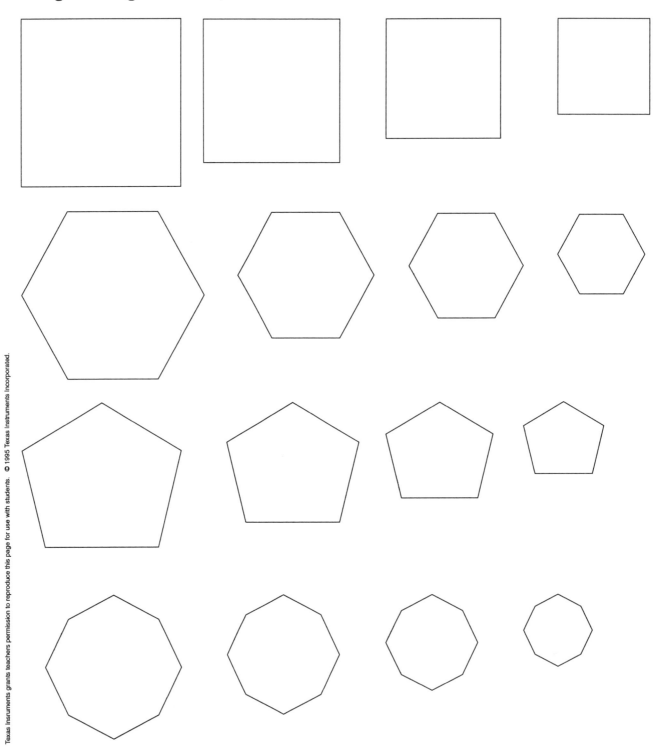

Predicting π

Math Concepts
- fractions
- decimals
- linear measure
- ratio
- proportion
- circles
- diameter
- circumference
- similarity

Materials
- TI-108, Math Mate™, Math Explorer™
- **Predicting π** recording sheets
- rulers, meter sticks, string, compasses, calipers
- pencils

Overview

Students will use linear measurement and calculators to discover the existence of π, the constant ratio between the circumference and the diameter of a circle.

Introduction

> It would be helpful to complete the **Ratios in Regular Polygons** activity on page 84 before beginning this activity.

1. Have students predict whether the distance around a soft drink can is less than, about the same as, or greater than its height. Have them graph their predictions in a class graph.

2. Have students use a piece of string to measure around the can and to measure its height. Have them compare the two lengths. Discuss the differences between the students' predictions and the actual measurements.

3. Then have students identify several circular objects in the classroom, on the school grounds, or at home.

   Note: Students might even collect objects to bring to class.

4. Have students select a tool to measure the circumference and diameter of each circle, and record these measurements on the recording sheet.

5. Have students use a compass to draw several different circles and record their circumferences and diameters on the recording sheet.

6. Have students look for patterns in their data and make conjectures about why the patterns might exist.

Cómo predecir π

Conceptos matemáticos

- fracciones
- decimales
- medida lineal
- relación
- proporción
- círculos
- diámetro
- circunferencia
- similitud

Materiales

- TI-108, Math Mate™, Math Explorer™
- hojas de registro de **Cómo predecir** π
- reglas, varas de medición, cuerda, compases, compases de espesor
- lápices

Resumen

Los alumnos utilizarán la medición lineal y la calculadora para descubrir la existencia de π, la relación constante entre la circunferencia y el diámetro de un círculo.

Introducción

> Antes de comenzar con esta actividad, conviene completar el ejercicio de **Relaciones en polígonos regulares** de la página 84.

1. Haga que los alumnos predigan si la distancia alrededor de una lata de bebida es menor, similar o mayor que su altura. Que grafiquen sus predicciones en un gráfico.

2. Que los alumnos utilicen un pedazo de cuerda para medir alrededor de la lata y su altura. Que comparen las dos longitudes. Analice las diferencias entre las predicciones de los estudiantes y las medidas reales.

3. Después, que identifiquen varios objetos circulares de la sala de clases, del patio de la escuela o de su casa.

   Nota: los alumnos incluso podrían traer objetos a la clase.

4. Que los alumnos seleccionen una herramienta para medir la circunferencia y el diámetro de cada círculo, y que anoten estas medidas en la hoja de registro.

5. Que los alumnos usen un compás para dibujar varios círculos y anoten su circunferencia y su diámetro en la hoja de registro.

6. Que los alumnos busquen patrones en sus datos y formulen conjeturas acerca de por qué los patrones podrían existir.

Predicting π (continued)

Collecting and Organizing Data

While students generate data for the circles, ask questions such as:

- How are all of the circles alike?

- How are you measuring the diameters?

- How are you measuring the circumferences?

- How are the measurements you are making with circles different from the measurements you made with the regular polygons (refer to **Ratios in Regular Polygons** on page 84) ? How are they alike?

- Does it matter if you measure in inches or centimeters? Why or why not? (see **Do Centimeters Make Me Taller?** on page 74).

- What patterns do you see?

- Why do you think those patterns are occurring?

How are you using the calculator to help you look for patterns?

How can you judge that what you see on your calculator is reasonable?

How can you use the calculator and the patterns you see to help you predict measurements of diameters or circumferences?

Analyzing Data and Drawing Conclusions

After students have made and compared several sets of measurements, have them discuss their results as a whole group. Ask questions such as:

- Did your data turn out exactly like everyone else's? Why or why not?

- What patterns do you see in your data?

- How are these patterns like the ones in the **Ratios in Regular Polygons** activity (page 84)? How are they different?

- How are the circumferences and the diameters of the circles related to each other?

- How is this relationship like the ones you found in the **Ratios in Regular Polygons** activity (page 84)?

- From the patterns in your data, what conclusions can you make about the number π, which represents the constant ratio between the circumference and diameter of a circle?

What operations or keys did you use on the calculator to help you find patterns in this activity? Why did you choose those operations or keys?

How did you determine if your calculator results were reasonable?

Cómo predecir π (continuación)

Cómo reunir y organizar los datos

Mientras los alumnos generan los datos de los círculos, haga las preguntas siguientes:

- ¿En qué se parecen todos los círculos?

- ¿Cómo están midiendo los diámetros?

- ¿Cómo están midiendo las circunferencias?

- ¿En que difieren las mediciones que están haciendo de los círculos y las mediciones hechas con los polígonos regulares (véa la sección **Relaciones en polígonos regulares** de la página 84)? ¿En qué se parecen?

- ¿Tiene alguna importancia si miden en pulgadas o en centímetros? ¿Por qué? (ver la sección **¿Los centímetros me hacen más alto?,** en la página 74).

- ¿Qué patrones observan?

- ¿Por qué piensan que se producen estos patrones?

⊞ ¿Cómo están usando la calculadora para buscar patrones?

⊞ ¿Cómo pueden juzgar si lo que ven en la calculadora es razonable?

⊞ ¿Cómo pueden utilizar la calculadora y los patrones que ven para predecir medidas de diámetros o circunferencias?

Cómo analizar los datos y sacar conclusiones

Después de que los alumnos elaboren y comparen varios grupos de medidas, pídales que analicen sus resultados como un solo grupo, usando las preguntas siguientes:

- ¿Sus datos resultaron exactamente iguales a los de los demás? ¿Por qué?

- ¿Qué patrones observan en sus datos?

- ¿En qué se parecen estos patrones a los de la actividad **Relaciones en polígonos regulares** (página 84)? ¿En qué se diferencian?

- ¿Cómo se relacionan las circunferencias y los diámetros de los círculos?

- ¿En qué se parece esta relación a las observadas en la actividad de **Relaciones en polígonos regulares** (página 84)?

- A partir de los patrones de sus datos, ¿qué conclusiones pueden sacar acerca del número π, que representa la relación constante entre la circunferencia y el diámetro de un círculo?

⊞ ¿Qué operaciones o teclas usaron en la calculadora para encontrar patrones en esta actividad? ¿Por qué escogieron esas operaciones y teclas?

⊞ ¿Cómo determinaron si los resultados de la calculadora eran razonables?

Predicting π (continued)

Analyzing Data and Drawing Conclusions (continued)

- Why do you think this ratio was given the name "Pi"?

- How can the knowledge of this constant ratio π be used?

- Do you think the distance around a tennis ball container is greater than, about the same as, or less than its height? Why?

Continuing the Investigation

Have students research the history of the development of the numerical value of π.

Cómo predecir π (continuación)

Cómo analizar los datos y sacar conclusiones (continuación)

- ¿Por qué piensan que a esta relación se le dio el nombre de "Pi"?

- ¿Cómo se puede usar el conocimiento de esta relación constante π?

- ¿Piensan que la distancia alrededor de un envase de pelotas de tenis es mayor, similar o menor que su altura? ¿Por qué?

Cómo continuar la investigación

Los alumnos deben investigar la historia evolutiva del valor numérico de π.

Name:

Predicting π
Recording Sheet

Collecting and Organizing Data

Object	Measure of Circumference (C)	Measure of Diameter (D)	Ratio of C to D	Ratio in Decimal Form

Analyzing Data and Drawing Conclusions

If we know the length of the diameter of a circle, I can find its circumference by:

If we know the length of the circumference of a circle, I can find the length of its diameter by:

Questions we thought of while we were doing this activity:

Nombre:

Cómo predecir π
Hoja de registro

Cómo reunir y organizar los datos

Objeto	Medida de la circunferencia (C)	Medida del diámetro (D)	Relación entre C y D	Relación en forma decimal

Cómo analizar los datos y sacar conclusiones

Si conocemos la longitud del diámetro de un círculo, puedo encontrar su circunferencia mediante el procedimiento siguiente:

Si conocemos la longitud de la circunferencia de un círculo, puedo encontrar la longitud de su diámetro mediante el procedimiento siguiente:

Algunas preguntas que surgieron mientras realizábamos esta actividad:

Levels
2 & 3

Activities:

1 Spin Me Along

2 Tiles in a Bag

3 Does One = One?

4 Number Cube Sums

5 Analyzing Number Cube Sums

6 Picturing Probabilities of Number Cube Sums

7 An Average Lunch?

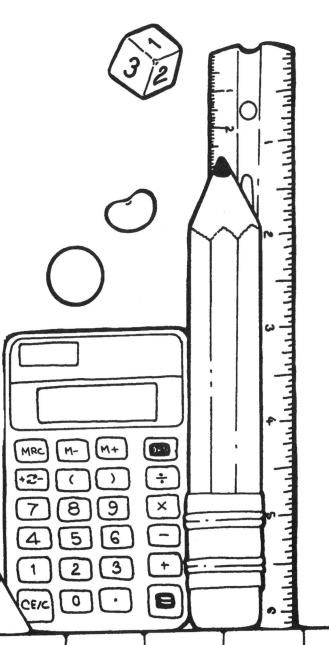

Spin Me Along

Math Concepts

- whole numbers
- fractions
- decimals
- area
- sample space
- probability

Materials

- Math Explorer™
- **Spin Me Along** recording sheets
- small paper clips
- pencils

Overview

Students will explore probability and patterns in fractions and decimals by spinning three spinners and recording and analyzing the results.

Introduction

1. Have students discuss Spinner A on page 97. Ask questions such as: What do you think will happen when you spin this spinner? Why do you think that? If we all spin Spinner A once and display our results in a bar graph, what do you think it will look like? Have students make similar predictions for Spinners B and C.

2. Make a simple spinning device by bending out one end of a paper clip. Then place the tip of a pencil through the curve at the end of the paper clip and onto the center of one of the spinner circles provided on page 97.

3. Model the **Spin Me Along** activity for students on the overhead projector using Spinner A on page 97. Tally the results for the first ten spins.

4. Show students how to record the fractional part of the spins for 1 and 2.

   Example: Out of ten spins, 1 comes up four times (4/10) and 2 comes up six times (6/10). Show students how to use $\boxed{F \supset D}$ on the calculator to change each fraction to a decimal.

Results

	Result of 1	Result of 2
Tally of Results *After 10 spins*	𝖳𝖧𝖫	𝖳𝖧𝖫 /
Fractions	4/10	6/10
Decimals	0.4	0.6
Tally of Results *After 10 spins*	𝖳𝖧𝖫	𝖳𝖧𝖫
Fractions	9/20	11/20
Decimals	0.45	0.55

Hazme girar

Conceptos matemáticos

- números enteros
- fracciones
- decimales
- superficie
- espacio muestreo
- probabilidad

Materiales

- Math Explorer™
- hojas de registro de **Hazme girar**
- clips para papel pequeños
- lápices

Resumen

Los alumnos examinarán la probabilidad y los patrones en fracciones y decimales, haciendo girar tres giradores y anotando y analizando los resultados.

Introducción

1. Haga que los alumnos analicen el Girador A de la página 97. Haga las preguntas siguientes: ¿qué piensan que sucedería si hacen girar este girador? ¿Por qué piensan eso? Si todos hacemos girar el Girador A una vez y ponemos nuestros resultados en un gráfico de barras, ¿cómo piensan que se vería? Que los alumnos hagan predicciones similares para los giradores B y C.

2. Construya un dispositivo giratorio simplemente doblando un extremo de un clip para papel. Después, coloque la punta de un lápiz a través de la curva en el extremo del clip y sobre el centro de uno de los círculos giratorios que vienen en la página 97.

3. Represente la actividad de **Hazme girar** en el proyector de transparencias, usando el Girador A de la página 97. Haga una cuenta de los resultados para los primeros diez giros.

4. Muestre a los alumnos cómo registrar la parte fraccional de las frecuencias de 1 y 2.

   Ejemplo: de diez veces, 1 salió cuatro veces (4/10) y 2 salió seis (6/10). Muestre a los alumnos cómo usar $\boxed{\text{F⊃D}}$ en la calculadora para cambiar cada fracción a un decimal.

Resultados

	Resultado de 1	Resultado de 2
Cuenta de Resultados *Después de 10 veces*	卌	卌 /
Fracciones	6/10	0.6
Decimales	0.4	4/10
Cuenta de Resultados *Después de 20 veces*	卌	卌
Fracciones	9/20	11/20
Decimales	0.45	0.55

Spin Me Along (continued)

Introduction (continued)

5. Have students spin Spinner A 40 times to collect data, recording the fractions and decimals after each 10 spins. Then have students spin and collect data in the same way for Spinners B and C.

6. As a whole class, compile the data into one class chart for each spinner. Then have students analyze the class data and write about what they notice.

Collecting and Organizing Data

While students are collecting data and recording the fractions and decimals, ask questions such as:

- Is each number equally likely to occur on each of the spinners? Why or why not?

- Which numbers do you think are more likely to occur than others? Why do you think that? Does the data you are collecting seem to support your ideas?

- How are you deciding which fraction to use to describe each outcome?

- What is the "whole" to which the fractions and decimals are referring?

- Do you see any patterns in the fractions and decimals you are recording?

How are you using the calculator to help you in this problem?

How can you use the F◻D key to compare fractions and decimals?

Would you want to use the INT÷ key to compare fractions and decimals? Why or why not?

How can you use the ÷ key to compare fractions and decimals?

Analyzing Data and Drawing Conclusions

After students have collected their data and added their information to a class chart, have them discuss the results as a whole group. Ask questions such as:

- What information did you use to predict which number would come up most often on each of the spinners?

- Are each of the numbers equally likely to come up on all three spinners? Why or why not?

When you change a fraction to a decimal and then back to a fraction again, sometimes you can use the F◻D key over and over and sometimes you have to switch to the x◻y key. Why do you think this happens?

Did you use the F◻D key to compare fractions and decimals? Why or why not?

Hazme girar (continuación)

Introducción (continuación)

5. Que los alumnos hagan girar el Girador A 40 veces para reunir datos y que registren las partes fraccionales y los decimales después de cada 10 giros. Después, que los alumnos hagan girar y reúnan datos de igual manera para los giradores B y C.

6. Como un solo grupo, recopile los datos en un cuadro general de todo el curso para cada girador. Después, que los alumnos analicen los datos generales y escriban sobre sus observaciones.

Cómo reunir y organizar los datos

Mientras los alumnos reúnen datos y registran las fracciones y los decimales, haga las preguntas siguientes:

- ¿Cada número tiene las mismas probabilidades de salir en cada uno de los giradores? ¿Por qué?

- ¿Qué números piensan que tienen mayor probabilidad de salir que otros? ¿Por qué piensan eso? ¿Los datos que están reuniendo respaldan sus ideas?

- ¿Cómo están decidiendo qué fracción usar para describir cada resultado?

- ¿Cuál es el "entero" al que se refieren las fracciones y los decimales?

- ¿Observan algún patrón en las fracciones y los decimales que están registrando?

▣ ¿Cómo están usando la calculadora en este problema?

▣ ¿Cómo pueden usar la tecla F⊂D para comparar fracciones y decimales?

▣ ¿Usarían la tecla INT÷ para comparar fracciones y decimales? ¿Por qué?

▣ ¿Cómo pueden usar la tecla ÷ para comparar fracciones y decimales?

Cómo analizar los datos y sacar conclusiones

Después de que los alumnos reúnan sus datos y agreguen su información al gráfico general, pídales que analicen los resultados. Haga las preguntas siguientes:

- ¿Qué información utilizaron para predecir qué número aparecería más frecuentemente en cada uno de los giradores?

- ¿Cada número tiene las mismas probabilidades de salir en los tres giradores? ¿Por qué?

▣ Cuando transforman una fracción en un decimal y luego vuelven a la fracción, a veces pueden usar la tecla F⊂D repetidamente y en otras ocasiones tienen que pasar a la tecla x⊂y. ¿Por qué creen que ocurre esto?

▣ ¿Utilizaron la tecla F⊂D para comparar fracciones y decimales? ¿Por qué?

Spin Me Along (continued)

Analyzing Data and Drawing Conclusions (continued)

- How did your individual results compare to the class results for each spinner?

- How could you describe the patterns in the fractions and decimals?

- What if you changed the sizes of the sections on the spinners? How do you think it would change your results?

Did you use the ÷ key to compare fractions and decimals? Why or why not?

Did you use the INT÷ key to compare fractions and decimals? Why or why not?

Continuing the Investigation

Have students:

- Change the sizes of the sections on the spinners, predict how the likelihood of the outcomes will change, and collect data to compare to their predictions.

- Invent a spinner that they think will produce a given set of results and collect data to compare to their predictions.

  Example: A spinner where 1 will come up half as often as 2 or a spinner where 2 will come up four times as often as 1.

© 1995 Texas Instruments Incorporated. ™ Trademark of Texas Instruments Incorporated.

Hazme girar (continuación)

Cómo analizar los datos y sacar conclusiones (continuación)

- ¿Cómo se comparaban sus resultados individuales con los de la clase en cada girador?

- ¿Cómo podrían describir los patrones en las fracciones y los decimales?

- ¿Qué sucedería si cambiaran los tamaños de las secciones en los giradores? ¿Cómo creen que cambiarían sus resultados?

▦ ¿Utilizaron la tecla ÷ para comparar fracciones y decimales? ¿Por qué?

▦ ¿Usaron la tecla INT÷ para comparar fracciones y decimales? ¿Por qué?

Cómo continuar la investigación

Los alumnos deben:

- Cambiar el tamaño de las secciones en los giradores, predecir cómo cambiarán las probabilidades de los resultados y reunir datos para compararlos con sus predicciones?

- Inventar un girador que piensan que producirá un conjunto determinado de resultados y reunir datos para compararlos con sus predicciones.

  Ejemplo: un girador en que 1 aparezca la mitad de veces que 2 o uno en que 2 aparezca cuatro veces más que 1.

Name:

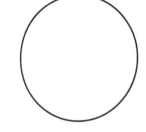

Spin Me Along
Recording Sheet

Collecting and Organizing Data

Results

	Result of 1	Result of 2
Tally of Results *After 10 spins*	_____	_____
Fractions	_____	_____
Decimals	_____	_____
Tally of Results *After 20 spins*	_____	_____
Fractions	_____	_____
Decimals	_____	_____
Tally of Results *After 30 spins*	_____	_____
Fractions	_____	_____
Decimals	_____	_____
Tally of Results *After 40 spins*	_____	_____
Fractions	_____	_____
Decimals	_____	_____

Analyzing Data and Drawing Conclusions

Write about the information that you gathered.

Nombre:

Hazme girar
Hoja de registro

Cómo reunir y organizar los datos

Resultados

	Resultado de 1	Resultado de 2
Cuenta de resultados *Después de 10 veces*	_____	_____
Fracciones	_____	_____
Decimales	_____	_____
Cuenta de resultados *Después de 20 veces*	_____	_____
Fracciones	_____	_____
Decimales	_____	_____
Cuenta de resultados *Después de 30 veces*	_____	_____
Fracciones	_____	_____
Decimales	_____	_____
Cuenta de resultados *Después de 40 veces*	_____	_____
Fracciones	_____	_____
Decimales	_____	_____

Cómo analizar los datos y sacar conclusiones

Escriban acerca de la información que reunieron.

Spin Me Along

Spinners

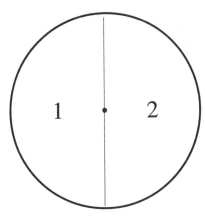

Spinner A

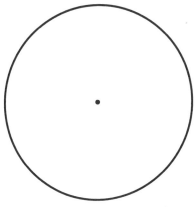

Spinner _____

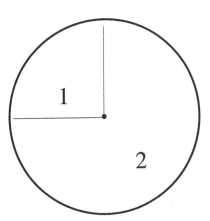

Spinner B

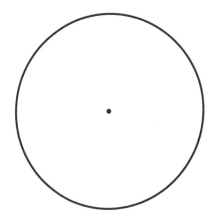

Spinner _____

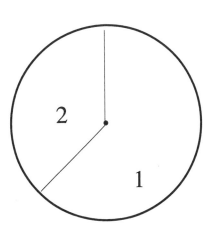

Spinner C

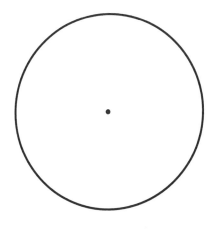

Spinner _____

Hazme girar

Giradores

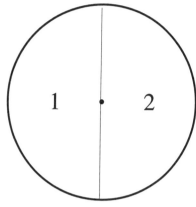

Girador A

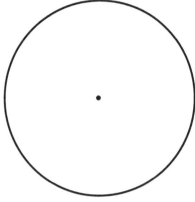

Girador _____

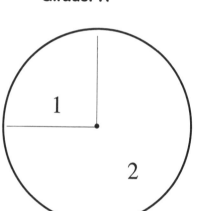

Girador B

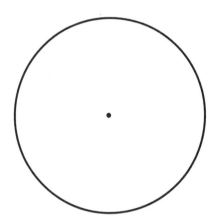

Girador _____

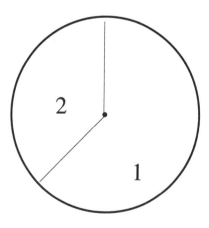

Girador C

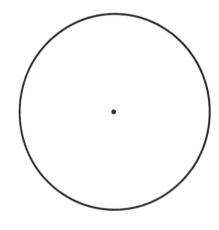

Girador _____

Tiles in a Bag

Math Concepts

- whole numbers
- fractions
- decimals
- sample space
- probability

Materials

- Math Explorer™
- **Tiles in a Bag** recording sheets
- tiles in two different colors
- small lunch bags
- pencils

Overview

Students will explore probability and patterns in fractions and decimals by drawing tiles out of a bag and recording and analyzing the results.

Introduction

1. Prepare a "mystery bag" containing ten tiles of two different colors; for example, six red and four blue. Ask students to guess how many tiles of each of the two different colors are in the bag. Have students draw out one tile, record its color, and replace it in the bag. After ten such trials, ask students if they wish to change their guesses.

2. Discuss with students how to record the fractions that represent the draws for each color.

   Example: From ten draws, 3 tiles are blue (3/10) and 7 tiles are red (7/10). Show students how to use $\boxed{F \subset D}$ on the calculator to change each fraction to a decimal.

Colors	red	blue
Tally of Results	7	3
Fractions	7/10	3/10
Decimals	0.7	0.3
After 10 draws *we think there are:*	7 out of 10	3 out of 10

3. Have a student draw 10 more times from the bag just as before, recording in fraction and decimal form the number of tiles drawn of each color. Give students another chance to change their predictions based on the new information.

   Note: This time the denominator of the fraction will be 20.

4. Have students go through the same process two more times until a total of 40 trials have been made.

5. Then ask students to make final predictions about the tiles in the bag.

Las fichas en la bolsa

Conceptos matemáticos

- números enteros
- fracciones
- decimales
- espacio muestreo
- probabilidad

Materiales

- Math Explorer™
- hojas de registro de **Las fichas en la bolsa**
- fichas de dos colores diferentes
- bolsas pequeñas
- lápices

Resumen

Los alumnos examinarán las probabilidades y los patrones en fracciones y decimales, sacando fichas de una bolsa y registrando y analizando los resultados.

Introducción

1. Muestre a los alumnos una "bolsa misteriosa" y pídales que adivinen cuántas fichas de dos colores diferentes hay dentro. Que los alumnos saquen una ficha, anoten el color y vuelvan a colocarla en la bolsa. Después de diez intentos, pregunte a los alumnos si desean cambiar sus pronósticos.

2. Analice con los alumnos cómo anotar la parte fraccional del total de turnos que correspondió a cada color.

   Ejemplo: de diez turnos, 3 fichas son azules (3/10) y 7 son rojas (7/10). Muestre a los alumnos cómo usar $\boxed{F\subset D}$ en la calculadora para transformar cada fracción en un decimal.

Colores	*rojo*	*azul*
Cuenta de resultados	*7*	*3*
Fracciones	*7/10*	*3/10*
Decimales	*0.7*	*0.3*
Después de 10 turnos pensamos que hay:	*7* de 10	*3* de 10

3. Que un alumno saque 10 fichas más de la bolsa, igual que las veces anteriores, registrando en forma fraccional y decimal el número de fichas de cada color. Dé a los alumnos otra oportunidad de cambiar sus predicciones sobre la base de la información nueva.

   Nota: esta vez el denominador de la fracción será 20.

4. Que los alumnos pasen por el mismo proceso dos veces más, hasta llegar a un total de 40 intentos.

5. Después, pida a los alumnos que hagan predicciones finales acerca de las fichas en la bolsa.

Tiles in a Bag (continued)

Introduction (continued)

6. Have students analyze the patterns in the fractions and decimals that they recorded as of their results after each group of ten draws.

7. Give each group of students a new bag with a different combination of tiles in two colors. Have them repeat the process with their new bags and write about their discoveries.

Collecting and Organizing Data

While students are collecting data and recording the fractions and decimals, ask questions such as:

- What information are you using to make your predictions each time?

- Have your results changed or remained the same after each ten draws? What do you think affects your results?

- How are you deciding which fraction to use to describe each color?

- What is the "whole" to which the fractions and decimals are referring?

- Do you see any patterns in the fractions and decimals that you are recording?

🖩 How are you using the calculator to help you?

🖩 How can you use [F⊃D] to compare fractions and decimals?

🖩 How can you use [÷] to compare fractions and decimals?

🖩 Would you want to use [INT÷] to compare fractions and decimals? Why or why not?

Analyzing Data and Drawing Conclusions

After students have collected their data, have them discuss the results as a whole group. Ask questions such as:

- What information did you use to predict which combination of colored tiles is in the bag?

- Is each of the colors equally likely to occur each time you draw a tile? Why or why not?

- Look at each set of ten draws. Is any set exactly the same as what you discovered was actually in the bag? How many sets were not the same? How can you explain the differences?

🖩 When you change a fraction to a decimal and then back to a fraction again, sometimes you can use [F⊃D] over and over and sometimes you have to switch to [x⊃y]. Why do you think this happens?

🖩 Did you use [F⊃D] to compare fractions and decimals? Why or why not?

© 1995 Texas Instruments Incorporated. ™ Trademark of Texas Instruments Incorporated.

Las fichas en la bolsa (continuación)

Introducción (continuación)

6. Que los alumnos analicen los patrones en los registros fraccionales y decimales de sus resultados después de cada grupo de diez turnos.

7. Entregue a cada grupo de alumnos una bolsa nueva con una combinación distinta de fichas de dos colores. Que repitan el proceso con las bolsas nuevas y escriban acerca de sus hallazgos.

Cómo reunir y organizar los datos

Mientras los alumnos reúnen datos y anotan las fracciones y los decimales, haga las preguntas siguientes:

- ¿Qué información están usando para hacer sus predicciones cada vez?

- ¿Sus resultados cambiaron o permanecieron iguales después de cada diez turnos? ¿Qué piensan que afecta sus resultados?

- ¿Cómo deciden qué fracción usar para describir cada resultado?

- ¿Cuál es el "entero" al que se refieren las fracciones y los decimales?

- ¿Observan algún patrón en las fracciones y decimales que están registrando?

⊞ ¿Cómo están usando la calculadora?

⊞ ¿Cómo pueden usar F⊂D para comparar fracciones y decimales?

⊞ ¿Cómo pueden usar ÷ para comparar fracciones y decimales?

⊞ ¿Usarían INT÷ para comparar fracciones y decimales? ¿Por qué?

Cómo analizar los datos y sacar conclusiones

Después de que los alumnos reúnan sus datos, pídales que analicen los resultados. Haga las preguntas siguientes:

- ¿Qué información utilizaron para predecir qué combinación de fichas de color hay en la bolsa?

- ¿Cada color tiene las mismas probabilidades de salir cada vez que sacan una ficha? ¿Por qué?

- Examinen cada grupo de diez turnos. ¿Algún grupo es exactamente igual a lo que hallaron realmente en la bolsa? ¿Cuántos grupos no eran iguales? ¿Cómo pueden explicar las diferencias?

⊞ Cuándo transforman una fracción en un decimal y luego vuelven a fracción, a veces pueden usar F⊂D repetidamente, y en otras ocasiones tienen que pasar a x⊂y. ¿Por qué creen que sucede esto?

⊞ ¿Utilizaron F⊂D para comparar fracciones y decimales? ¿Por qué?

Tiles in a Bag (continued)

Analyzing Data and Drawing Conclusions (continued)

- How could you describe the patterns in the fractions and decimals?

- What if you changed the number of tiles in the bag? How do you think your results would change?

⊞ Did you use ÷ to compare fractions and decimals? Why or why not?

⊞ Did you use INT÷ to compare fractions and decimals? Why or why not?

Continuing the Investigation

Have students:

- Change the combinations of tiles in their bags, predict how the likelihood of the outcomes will change, and collect data to compare to their predictions.

- Put together a combination of tiles that will produce a given set of results and collect data to compare to their predictions.

Las fichas en la bolsa (continuación)

Cómo analizar los datos y sacar conclusiones (continuación)

- ¿Cómo podrían describir los patrones en las fracciones y los decimales?

- ¿Qué sucedería si cambiaran el número de fichas en la bolsa? ¿Cómo piensan que cambiarían sus resultados?

🖩 ¿Utilizaron ÷ para comparar fracciones y decimales? ¿Por qué?

🖩 ¿Querían usar INT÷ para comparar fracciones y decimales? ¿Por qué?

Cómo continuar la investigación

Los alumnos deben:

- Cambiar las combinaciones de fichas en las bolsas, predecir cómo cambiarán las probabilidades de los resultados y reunir datos para compararlos con sus predicciones.

- Elaborar una combinación de fichas que producirá un conjunto determinado de resultados y reunir datos para compararlos con sus predicciones.

Name:

Tiles in a Bag
Recording Sheet

Collecting and Organizing Data

Colors _____   _____

Tally of Results _____   _____

Fractions _____   _____

Decimals _____   _____
*After 10 draws,
we think there are:* _____out of 10   _____out of 10

Tally of Results _____   _____

Fractions _____   _____

Decimals _____   _____
*After 20 draws,
we think there are:* _____out of 20   _____out of 20

Tally of Results _____   _____

Fractions _____   _____

Decimals _____   _____
*After 30 draws,
we think there are:* _____out of 30   _____out of 30

Tally of Results _____   _____

Fractions _____   _____

Decimals _____   _____
*After 40 draws,
we think there are:* _____out of 40   _____out of 40

Analyzing Data and Drawing Conclusions

Write about what your information shows you on the back of this sheet.

Nombre:

Las fichas en la bolsa
Hoja de registro

Cómo reunir y organizar los datos

Colores

_____   _____

_____

Cuenta de resultados   _____   _____

Fracciones   _____   _____

Decimales   _____   _____
Después de 10 turnos,
pensamos que hay: _____de 10   _____de 10

Cuenta de resultados   _____   _____

Fracciones   _____   _____

Decimales   _____   _____
Después de 20 turnos,
pensamos que hay: _____de 20   _____de 20

Cuenta de resultados   _____   _____

Fracciones   _____   _____

Decimales   _____   _____
Después de 30 turnos,
pensamos que hay: _____de 30   _____de 30

Cuenta de resultados   _____   _____

Fracciones   _____   _____

Decimales   _____   _____
Después de 40 turnos,
pensamos que hay: _____de 40   _____de 40

Cómo analizar los datos y sacar conclusiones

Escriban acerca de lo que la información les indica en el reverso de esta hoja.

Does One = One?

Math Concepts

- whole numbers
- fractions
- decimals
- rounding
- sample space
- probability

Materials

- Math Explorer™
- number cube (regular die)
- polyhedral dice
- different colored linking cubes or centimeter cubes
- blank spinners
- **Does One = One?** recording sheets
- pencils

Overview

Students will extend their understanding of probability by identifying probabilities of equally-likely events occurring in a sample space. They will compare sums of fractional probabilities and their decimal representations on the calculator to develop awareness that a fraction and its decimal representation on the calculator are "close," but not necessarily equal.

Introduction

1. Have students look at a regular six-sided die and list all the possible outcomes for rolling it one time. This list represents the sample space for the experiment of rolling a six-sided die one time.

2. Discuss with students the fact that each of the outcomes is equally likely to happen when the die is rolled. Have students create statements using fractional probabilities to describe this "equally-likely" relationship.

   Example: You have 1/6 of a chance of rolling a 1 each time you roll the die.

3. Have students discuss the probability of rolling either a 1, 2, 3, 4, 5, or 6. Since it is certain that one of these numbers will show up on a roll, the probability is 1 and represents the probability of the entire sample space.

4. There is a way to demonstrate for students that the probability of the sample space is 1. Add together all the probabilities of the equally-likely outcomes. Use the repeated addition sequence $\boxed{+}$ **1** $\boxed{/}$ **6** $\boxed{=}$ $\boxed{=}$ $\boxed{=}$ $\boxed{=}$ $\boxed{=}$ $\boxed{=}$ to show that $1/6 + 1/6 + 1/6 + 1/6 + 1/6 + 1/6 = 6/6 = 1$.

   Note: Students can use the $\boxed{\text{Simp}}$ key to simplify the sum to 1.

¿Es uno = uno?

Conceptos matemáticos

- números enteros
- fracciones
- decimales
- redondeo
- espacio muestreo
- probabilidad

Materiales

- Math Explorer™
- dados poliédricos
- cubos de enlace de diferentes colores o cubos de un centímetro de lado
- giradores en blanco
- hojas de registro de ¿**Es uno = uno?**
- lápices

Resumen

Los alumnos aumentarán su comprensión del concepto de probabilidad, identificando probabilidades de hechos igualmente probables que ocurren en un espacio muestreo. Compararán las sumas de probabilidades fraccionales y sus representaciones decimales en la calculadora para desarrollar sus conocimientos de que una fracción y su representación decimal en la calculadora son "aproximadas", pero no necesariamente iguales.

Introducción

1. Pida a los alumnos que observen un dado de seis lados y anoten todos los resultados posibles si lo tiraran una vez. Esta lista representa el espacio muestreo para el experimento de tirar un dado de seis lados una vez.

2. Analice con los alumnos el hecho de que cada resultado es igualmente probable cuando se tira el dado. Que los alumnos formulen afirmaciones usando probabilidades fraccionales para describir esta relación "igualmente probable".

   Ejemplo: tienen 1/6 de posibilidad de sacar un 1 cada vez que lanzan el dado.

3. Que los alumnos analicen la probabilidad de sacar 1, 2, 3, 4, 5 ó 6. Como es seguro que saldrá uno de estos números al tirar el dado, la probabilidad es 1 y representa la probabilidad de todo el espacio muestreo.

4. Hay una forma de demostrarles a los alumnos que la probabilidad de espacio muestreo es 1. Sume todas las probabilidades de los resultados igualmente probables. Use la secuencia de suma repetida ⊞ **1** ⟋ **6** ＝＝＝＝＝＝ para mostrar que 1/6 + 1/6 + 1/6 + 1/6 + 1/6 + 1/6 = 6/6 = 1.

   Nota: los alumnos pueden usar la tecla ⟨Simp⟩ para simplificar la suma a 1.

Does One = One? (continued)

Introduction (continued)

5. Ask students: Do you think it matters whether the probability is described with a fraction or a decimal? Let students explore with the calculator. Have them use the [F⊃D] key to change 1/6 to a decimal. Then have them enter the decimal and use it in a repeated addition sequence just as you demonstrated to them with the fraction form ([+] **0.1666667** [=] [=] [=] [=] [=] [=]) to show that 0.1666667 added together six times is 1.0000002.

   Note: Different calculators handle repeating fractions differently. Some round, and some just stop when the display is full.

6. Have students record the information from their explorations with 1/6 on the recording sheet, discuss what they see happening, and make some conjectures about why it is happening.

Description of Experiment Number of Equally-Likely Outcomes in Sample Space		Probability of Each Outcome		Sum of Probabilities of All Outcomes in Sample Space	
		Fraction	[F⊃D]	Fraction	[F⊃D]
Rolling a die	6	1/6	0.1666667	6/6	1.0000002

7. Have students think of some other experiments (rolling other shapes of polyhedral dice, spinning spinners, or drawing cubes from a bag) where each outcome has an equally-likely chance of happening.

   Example: Spinning a spinner with five congruent sections that each have a different label has a sample space of five equally-likely outcomes.

   Note: Although the students are working here with theoretical probabilities and don't need to be actually rolling dice or spinning spinners, it may be helpful for them to have these objects around to see.

8. Ask students to make some conjectures about what is happening when the calculator's decimal representations do not add up to 1 and then test their conjectures by exploring other examples of fractions and their decimal representations on the calculator.

9. Have students write about their discoveries.

© 1995 Texas Instruments Incorporated. ™ Trademark of Texas Instruments Incorporated.

¿Es uno = uno? (continuación)

Introducción (continuación)

5. Pregunta para los alumnos: ¿piensan que tiene alguna importancia si la probabilidad se describe con una fracción o con un decimal? Deje que los alumnos prueben con la calculadora. Pídales que utilicen la tecla $\boxed{\text{F}\supset\text{D}}$ para cambiar 1/6 a un decimal. Después, que ingresen el decimal y lo usen en una secuencia de suma repetida, de la misma manera en que usted les demostró con la forma de fracción ($\boxed{+}$ **0.1666667** $\boxed{=}\boxed{=}\boxed{=}\boxed{=}\boxed{=}\boxed{=}$) para demostrar que 0.1666667 sumado seis veces es 1.0000002.

   Nota: Las calculadoras manejan las fracciones repetidas de manera diferente. Algunas redondean y otras simplemente se detienen cuando el visor está lleno.

6. Pida a los alumnos que registren la información a partir de sus experiencias con 1/6 en la hoja de registro, analicen lo que observan y formulen algunas conjeturas acerca de por qué sucede esto:

Descripción del experimento	Probabilidad de cada resultado		Suma de las probabilidades de todos los resultados en el espacio muestrero	
Número de resultados igualmente probables en el espacio muestreo	Fracción	$\boxed{\text{F}\supset\text{D}}$	Fracción	$\boxed{\text{F}\supset\text{D}}$
Tirar un dado 6	1/6	0.1666667	6/6	1.0000002

7. Que los alumnos piensen en otros experimentos (tirar dados poliédricos con otras formas, hacer girar un girador, o sacar cubos de una bolsa) en que cada resultado tenga una posibilidad igualmente probable de salir.

   Ejemplo: al hacer girar un girador con cinco secciones congruentes y que cada una tenga una marca distinta, se tiene un espacio muestreo de cinco resultados igualmente probables.

   Nota: aunque los alumnos están trabajando aquí con probabilidades teóricas y no necesitan tirar realmente los dados o hacer girar los giradores, sería útil que tuvieran los objetos a su alcance para observarlos.

8. Pida a los alumnos que formulen algunas conjeturas acerca de lo que está sucediendo cuando las representaciones decimales de la calculadora no suman 1 y después que prueben sus conjeturas examinando otros ejemplos de fracciones y sus representaciones decimales en la calculadora.

9. Que los alumnos escriban acerca de sus descubrimientos.

Does One = One? (continued)

Collecting and Organizing Data

While students are examining their fractions and the decimal representations on the calculator, ask questions such as:

- How many equally-likely outcomes does this (die, spinner, bag of cubes) have?

- What fractional probability does each equally-likely outcome have?

- How does the calculator show this fraction as a decimal?

- If the fractions and decimal representations are equal to each other, what do you predict should happen when you add the decimal representations together? Try it and see.

- How can you explain the results you are getting?

- What is the same about the fractions whose decimal representations on the calculator add up to 1? How about the ones that do not add up to 1? How are they the same?

- How can you use your observations to make up some additional sample spaces and fractional probabilities to test?

🖩 How are you using the calculator to help you investigate this problem?

🖩 How do you know if the sum of the fractions is reasonable or not?

🖩 How do you know if the sum of the decimals is reasonable or not?

🖩 If F⊃D is supposed to give you decimals that are equivalent to the fractions, why are the sums not always the same?

Analyzing Data and Drawing Conclusions

After students have collected their data, have them discuss the results as a whole group. Ask questions such as:

- Report your observations. Did anyone notice anything else? Can you think of another way to describe what you are noticing?

- What fractions did you try? Why did you choose those fractions?

- What information did you use to make your conjectures?

- What examples did you use to test your conjecture? Did you find any counterexamples? How did you use your counterexamples to change your conjecture?

- When might this difference in sum matter? When might it not matter? When is "close" not close enough?

🖩 When you change a fraction to a decimal and then back to a fraction again, sometimes you can use F⊃D over and over and sometimes you have to switch to x⊃y. Why do you think this happens?

¿Es uno = uno? (continuación)

Cómo reunir y organizar los datos

Mientras los alumnos examinan sus fracciones y las representaciones decimales de la calculadora, haga las preguntas siguientes:

- ¿Cuántos resultados igualmente probables tiene este proceso (dado, girador, bolsa de cubos)?

- ¿Que probabilidad fraccional tiene cada resultado igualmente probable?

- ¿De qué manera muestra la calculadora esta fracción como decimal?

- Si las fracciones y representaciones decimales son iguales entre sí, ¿qué predicen que ocurriría si suman todas las representaciones decimales? Inténtenlo y vean.

- ¿Cómo pueden explicar los resultados que obtienen?

- ¿Cuál es el elemento común en las fracciones cuyas representaciones decimales en la calculadora suman 1? ¿Qué sucede con las que no suman 1? ¿En qué son iguales?

- ¿Cómo pueden usar sus observaciones para elaborar otros espacios muestreo y probabilidades fraccionales para comprobar?

▣ ¿Cómo están usando la calculadora para investigar este problema?

▣ ¿Cómo saben si la suma de las fracciones es razonable o no?

▣ ¿Cómo saben si la suma de los decimales es razonable o no?

▣ Si se supone que F◌D da decimales equivalentes a las fracciones, ¿por qué las sumas no son siempre las mismas?

Cómo analizar los datos y sacar conclusiones

Después de que los alumnos reúnan sus datos, pídales que analicen los resultados. Haga las preguntas siguientes:

- Hagan un informe de sus observaciones. ¿Alguien observó algo más? ¿Pueden pensar en otra forma de describir sus observaciones?

- ¿Qué fracciones probaron? ¿Por qué escogieron estas fracciones?

- ¿Qué información utilizaron para formular sus conjeturas?

- ¿Qué ejemplos utilizaron para probar su conjetura? ¿Encontraron algún contraejemplo? ¿Cómo utilizaron sus contraejemplos para cambiar su conjetura?

- ¿Cuándo tendría importancia esta diferencia en la suma? ¿Cuándo no tendría importancia? ¿Cuándo "lo suficientemente aproximado" no es lo suficientemente aproximado?

▣ Cuándo transforman una fracción en un decimal y luego vuelven a fracción, a veces pueden usar F◌D repetidamente, y en otras ocasiones tienen que pasar a x◌y. ¿Por qué creen que sucede esto?

Does One = One? (continued)

Continuing the Investigation

Have students:

- Make a class list of unit fractions (1/n, n ≠ 0) whose repeated addition sequences in decimal form on the calculator add up to 1 and a separate list of unit fractions whose decimal representations do not add up to 1. With the lists, post student conjectures, examples, and counterexamples.

- Choose pairs of fractions, use the calculator to find their sums in fraction form, and then use F⊂D to put the sum into decimal form. Students then change each of the addends to decimal form, add the decimals, and find the sum. Students should record the two sums, both in decimal form; discuss whether the sums are equal or not; and make conjectures about what might account for the differences.

¿Es uno = uno? (continuación)

Cómo continuar la investigación

Los alumnos deben:

- Hacer una lista general de las fracciones de unidades (1/n, n ≠ 0) cuyas secuencias de suma repetida en forma de decimal en la calculadora suman 1 y una lista separada de las fracciones de unidades cuyas representaciones decimales no suman 1. Con las listas, analice las conjeturas, ejemplos y contraejemplos de los alumnos.

- Escoger pares de fracciones, usar la calculadora para encontrar sus sumas en forma de fracción, y después usar F⊃D para poner la suma en forma decimal. A continuación, los alumnos cambian cada uno de los sumandos a forma decimal, suman los decimales y encuentra la suma. Los alumnos deben registrar las dos sumas en forma decimal, analizar si las sumas son iguales o no y formular conjeturas acerca de cuál sería la causa de las diferencias.

Name:

Does One = One?

Recording Sheet

Collecting and Organizing Data

Description of Experiment	Probability of Each Outcome		Sum of Probabilities of All Outcomes in Sample Space	
Number of Equally-Likely Outcomes in Sample Space	Fraction	F⊃D	Fraction	F⊃D

Analyzing Data and Drawing Conclusions

We think that the fractions and decimals both add up to 1 when:

We think that the fractions and decimals do not both add up to 1 when:

What this tells us about decimal representations on the calculator is:

Nombre:

¿Es uno = uno?

Hoja de registro

Cómo reunir y organizar los datos

Descripción del experimento		Probabilidad de cada resultado		Suma de las probabilidades de todos los resultados en el espacio muestrero	
Número de resultados igualmente probables en el espacio muestreo		Fracción	F⊃D	Fracción	F⊃D

Cómo analizar los datos y sacar conclusiones

Pensamos que las fracciones y los decimales suman 1 cuando:

Pensamos que las fracciones y los decimales no suman 1 cuando:

Lo que esto nos indica acerca de las representaciones decimales en la calculadora es lo siguiente:

Number Cube Sums

Math Concepts

- whole numbers
- fractions
- decimals
- sample space
- probability

Materials

- Math Explorer™
- **Number Cube Sums** recording sheets
- class graph
- number cubes
- pencils

Overview

Students will explore experimental probability and patterns in fractions and decimals by rolling two number cubes and recording and analyzing the sums that come up.

Introduction

1. Show students a pair of number cubes. Ask them: How many different sums are possible if you roll the number cubes and add the numbers that come up together?

   Examples: $1 + 1, 1 + 2, 1 + 3, 1 + 4, 1 + 5, 1 + 6, 2 + 1, 2 + 2, 2 + 3$, etc.

   Verify the suggested number of sums by analyzing the possible combinations that can be made with the numbers on two different number cubes.

2. Have students predict which sum will come up the most if the number cubes are rolled a large number of times.

3. Have all students roll the number cubes 50 times and tally the sums that occur. Then have them record their results in both fraction and decimal form.

   Example: If the sum 7 comes up 10 times out of 50 rolls, the tally in its fractional form is 10/50 (or 1/5) and 0.2 in its decimal form.

4. Ask students to analyze the results and decide whether to revise their predictions concerning which sum will come up the most often.

5. Ask each student to record his or her results on a class graph in which each tally represents more than one piece of data.

Sumas con los cubos con números

MATH EXPLORER

Conceptos matemáticos

- números enteros
- fracciones
- decimales
- espacio muestreo
- probabilidad

Materiales

- Math Explorer™
- hojas de registro de **Sumas con los cubos con números**
- gráfico general
- cubos con números
- lápices

Resumen

Los alumnos examinarán la probabilidad y los patrones en fracciones y decimales, tirando dos cubos con números y anotando y analizando las sumas que aparecen.

Introducción

1. Muestre a los alumnos un par de cubos con números. Pregúnteles lo siguiente: ¿cuántas sumas distintas son posibles, si tiran los cubos y suman los números que salen?

   Ejemplos: 1 + 1, 1 + 2, 1 + 3, 1 + 4, 1 + 5, 1 + 6, 2 + 1, 2 + 2, 2 + 3, etc.

   Verifique el número sugerido de sumas, analizando las combinaciones posibles que se pueden hacer con los números en dos cubos numerados distintos.

2. Que los alumnos predigan la suma que aparecerá con más frecuencia, si los cubos se tiran más veces.

3. Que todos los alumnos tiren los cubos 50 veces y que registren las sumas que se producen. Después, que anoten sus resultados en forma de fracción y de decimal.

   Ejemplo: si la suma 7 sale diez veces en 50 turnos, la cuenta en forma fraccional es 10/50 (o bien, 1/5) y 0.2 en su forma decimal.

4. Pida a los alumnos que analicen los resultados y decidan si van a revisar sus predicciones acerca de la suma que saldrá con mayor frecuencia.

5. Pídale a cada alumno que registre sus resultados en un gráfico general en el que cada cuenta representa más que un dato.

Number Cube Sums (continued)

Introduction (continued)

Example:

Tally of Results	Possible Sums	Fraction	Decimal
/	2	1/50	0.02
////	3	4/50	0.08
////	4	4/50	0.08
7HH ///	5	8/50	0.16
7HH 7HH ///	6	13/50	0.26
	•		
	•		
	•		

6. Have students record the class results using both fraction and decimal representations.

7. Have students analyze the patterns in the fractions and decimals used to record the results.

8. Ask students to write about their observations and discoveries.

Collecting and Organizing Data

While students are collecting data and recording the fractions and decimals, ask questions such as:

- What information are you using to make your predictions?

- Which sums seem to be coming up the most often? Why do you think they are?

- What do you think affects your results? Does "luck" have anything to do with the results you are getting? Why or why not?

- What information are you using to determine the fractions that represent how often each sum occurred?

- What is the "whole" to which the fractions and decimals are referring?

- Do you see any patterns in the fractions and decimals you are recording?

How are you using the calculator to help you?

How can you use F⊃D to compare fractions and decimals?

How can you use ÷ to compare fractions and decimals?

Would you want to use INT÷ to compare fractions and decimals? Why or why not?

Sumas con los cubos con números

(continuación)

Introducción (continuación)

Ejemplo:

Cuenta de resultados	Resultado posible	Fracción	Decimal
/	2	1/50	0.02
////	3	4/50	0.08
////	4	4/50	0.08
7HH ///	5	8/50	0.16
7HH 7HH ///	6	13/50	0.26
•			
•			
•			

6. Que los alumnos registren los resultados generales usando representaciones fraccionales y decimales.

7. Que los alumnos analicen los patrones en el registro fraccional y decimal de los resultados.

8. Pida a los alumnos que escriban acerca de sus observaciones y descubrimientos.

Cómo reunir y organizar los datos

Mientras los alumnos reúnen datos y registran las fracciones y los decimales, haga las preguntas siguientes:

- ¿Qué información están usando para hacer sus predicciones?

- ¿Qué sumas parecen salir con mayor frecuencia? ¿Por qué piensan que sucede eso?

- ¿Qué piensan que afecta los resultados? ¿La "suerte" tiene algo que ver con los resultados que están obteniendo? ¿Por qué?

- ¿Qué información están usando para determinar las fracciones que representan la frecuencia con que aparece cada suma?

- ¿Cuál es el entero al que se refieren las fracciones y los decimales?

- ¿Observan algún patrón en las fracciones y decimales que están registrando?

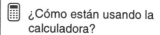 ¿Cómo están usando la calculadora?

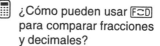 ¿Cómo pueden usar F⊂D para comparar fracciones y decimales?

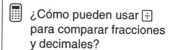 ¿Cómo pueden usar ÷ para comparar fracciones y decimales?

⊞ ¿Usarían INT÷ para comparar fracciones y decimales? ¿Por qué?

Number Cube Sums (continued)

Analyzing Data and Drawing Conclusions

After students have collected their data, have them discuss the results as a whole group. Ask questions such as:

- What information did you use to predict which sum would occur the most often?

- Is each of the sums equally likely to occur each time you roll the number cubes? Why or why not?

- Look at your set of 50 rolls on your recording sheet. Compare these results to the class results. How are they the same? How do they differ? How can you explain the differences?

- How could you describe the patterns in the fractions and decimals?

- How do the patterns differ when you compare your individual data with the data on the class graph? How are they the same?

- What does each tally represent on the class graph? How did you decide to represent data that did not come out to be a whole tally mark?

⊞ When you change a fraction to a decimal and then back to a fraction again, sometimes you can use F⊃D over and over and sometimes you have to switch to the x⊃y key. Why do you think this happens?

⊞ Did you use F⊃D to compare fractions and decimals? Why or why not?

⊞ How can you use ÷ to compare fractions and decimals?

⊞ Did you use ÷ to compare fractions and decimals? Why or why not?

⊞ Did you use INT÷ to compare fractions and decimals? Why or why not?

Continuing the Investigation

Have students use other polyhedral dice, predict how the likelihood of the outcomes will change, and collect data to compare to their predictions.

Sumas con los cubos con números

(continuación)

Cómo analizar los datos y sacar conclusiones

Después de que los alumnos reúnan sus datos, pídales que analicen los resultados. Haga las preguntas siguientes:

- ¿Qué información utilizaron para predecir qué suma se produciría con mayor frecuencia?

- ¿Cada suma tiene las mismas probabilidades de salir cada vez que lanzan los cubos con números? ¿Por qué?

- Observe el total de 50 turnos en su hoja de registro. Compare estos resultados con los de la clase. ¿En qué son iguales? ¿En qué se diferencian? ¿Cómo pueden explicar estas diferencias?

- ¿Cómo podrían describir los patrones en las fracciones y los decimales?

- ¿En qué se diferencian los patrones al comparar los datos individuales con los del gráfico general? ¿En qué son iguales?

- ¿Qué representa cada cuenta en el gráfico general? ¿Cómo decidieron representar los datos que no resultaron ser una marca de cuenta entera?

▦ Cuándo cambian una fracción por un decimal y luego vuelven a la fracción, a veces pueden usar la tecla F⊂D repetidamente y en otras ocasiones tienen que pasar a la tecla x⊂y. ¿Por qué creen que ocurre esto?

▦ ¿Utilizaron la tecla F⊂D para comparar fracciones y decimales? ¿Por qué?

▦ ¿Cómo pueden usar la tecla & para comparar fracciones y decimales?

▦ ¿Utilizaron ÷ para comparar fracciones y decimales? ¿Por qué?

▦ ¿Utilizaron INT÷ para comparar fracciones y decimales? ¿Por qué?

Cómo continuar la investigación

Que los alumnos usen otros dados poliédricos, predigan cómo cambiará la probabilidad de los resultados y reúnan datos para compararlos con sus predicciones?

Name:

Number Cube Sums
Recording Sheet

Collecting and Organizing Data

Tally of Results	Possible Sums	Fraction	Decimal
_____	2	_____	
_____	3	_____	
_____	4	_____	
_____	5	_____	
_____	6	_____	
_____	7	_____	
_____	8	_____	
_____	9	_____	
_____	10	_____	
_____	11	_____	
_____	12	_____	

Add your information to the class graph.

Analyzing Data and Drawing Conclusions

Write about what the information you gathered shows you.

Nombre:

Sumas con los cubos con números
Hoja de registro

Cómo reunir y organizar los datos

Cuenta de resultados	Resultado posible	Fracción	Decimal
_____	2	_____	
_____	3	_____	
_____	4	_____	
_____	5	_____	
_____	6	_____	
_____	7	_____	
_____	8	_____	
_____	9	_____	
_____	10	_____	
_____	11	_____	
_____	12	_____	

Agreguen su información al gráfico general.

Cómo analizar los datos y sacar conclusiones

Escriban acerca de lo que representa la información que reunieron.

Analyzing Number Cube Sums

MATH EXPLORER

Math Concepts
- whole numbers
- fractions
- decimals
- sample space
- probability

Materials
- Math Explorer™
- **Analyzing Number Cube Sums** recording sheets
- small group and class results from **Number Cube Sums**
- number cubes
- pencils

Overview

Students will extend their understanding of theoretical probability and patterns. Using number cubes, they will build awareness that a fraction and its decimal representation on the calculator are "close," but not necessarily equal.

Introduction

> The **Number Cube Sums** activity on page 107 should be completed before beginning this activity.

1. Have students refer to their fraction and decimal representations from **Number Cube Sums**. Ask students to summarize why they think the experimental probabilities came out the way they did.

2. Ask students to record all the ways of rolling two number cubes to get each possible sum of 2 through 12.

   Note: If students use number cubes of two different colors, it will become clear that a sum of 3 can be rolled in two different ways: 1 + 2 and 2 + 1.

3. Have students record the theoretical probability of each sum in both fraction and decimal form.

   Example: The probability of rolling a sum of 6 is 5/36 or approximately 0.1388889.

4. Have students investigate the sum of all the fractional probabilities and the sum of all the decimal representations found with the calculator.

5. Ask students to write about their observations and discoveries.

Collecting and Organizing Data

While students are recording the fractions and decimals for the probability of each sum, ask questions such as:

- Why are you using this fraction to describe the probability of rolling a __?

 How are you using the calculator to help you?

Cómo analizar las sumas con los cubos con números

Conceptos matemáticos

- números enteros
- espacio muestreo
- fracciones
- probabilidad
- decimales

Materiales

- Math Explorer™
- hojas de registro de **Cómo analizar las sumas con los cubos con números**
- resultados grupales y generales de la actividad de **Sumas con los cubos con números**
- cubos con números
- lápices

Resumen

Los alumnos aumentarán su comprensión de probabilidad y patrones. Al usar cubos con números, sabrán que una fracción y su representación decimal en la calculadora son "aproximadas", pero no necesariamente iguales.

Introducción

> Antes de empezar esta actividad, se debe completar el ejercicio de **Sumas con los cubos con números** de la página 107.

1. Para esta actividad, que los alumnos copien en sus hojas de registro las representaciones fraccionales y decimales de la sección **Sumas con los cubos con números**. Pídales que resuman por qué piensan que el experimento resultó así.

2. Pida a los alumnos que registren todas las formas de lanzar dos cubos con números para obtener cada resultado posible de 2 a 12.

   Nota: si los alumnos usan cubos con números de dos colores diferentes, quedará claro que una suma de 3 se puede obtener de dos maneras diferentes: 1 + 2 y 2 + 1.

3. Que los alumnos anoten la probabilidad de cada resultado, usando las representaciones fraccionales y decimales.

   Ejemplo: la probabilidad de obtener un 6 es 5/36 o aproximadamente 0,1388889.

4. Que los alumnos investiguen la suma de todas las probabilidades y la suma de todas las representaciones decimales encontradas con la calculadora.

5. Pida a los alumnos que escriban acerca de sus observaciones y descubrimientos.

Cómo reunir y organizar los datos

Mientras los alumnos registran las fracciones y decimales para la probabilidad de cada resultado, haga las preguntas siguientes:

- ¿Por qué están usando esta fracción para describir la probabilidad de obtener un ___?

¿Cómo están usando la calculadora?

Analyzing Number Cube Sums (continued)

Collecting and Organizing Data (continued)

- What is the "whole" to which the fractions and decimals are referring?

- What do you notice about the denominator of each of your fractions? What does the denominator represent? What does this numerator represent? How about this one?

- Do you notice any patterns developing in your table? How could you describe them?

- What do you notice about the sum of the fractions?

- What do you notice about the sum of the decimals?

How can you use F◁D to compare fractions and decimals?

How can you use ÷ to compare fractions and decimals?

Would you want to use INT÷ to compare fractions and decimals? Why or why not?

Analyzing Data and Drawing Conclusions

After students have collected their data, have them discuss the results as a whole group. Ask questions such as:

- How did using number cubes of two different colors help you verify what you were recording in your table?

- How does the information in your table compare to the results in **Number Cube Sums**?

- What was the sum of your fractional probabilities? What sums did other groups get? How can you explain this?

- What was the sum of your probabilities as the calculator showed them in decimal form? How do you explain the difference?

- Report your observations. How can you explain what you observed?

When you change a fraction to a decimal and then back to a fraction again, sometimes you can use F◁D over and over and sometimes you have to switch to x◁y. Why do you think this happens?

How can you explain what the calculator appears to be doing as it changes fractions to decimals?

Continuing the Investigation

Have students use other polyhedral dice and predict each outcome in the sample space. Have them perform a similar investigation with the fraction and decimal representations of the probability of each possible sum.

Cómo analizar las sumas con los cubos con números (continuación)

Cómo reunir y organizar los datos (continuación)

- ¿Cuál es el entero al que se refieren las fracciones y los decimales?

- ¿Qué observan acerca del denominador de cada una de sus fracciones? ¿Qué representa el denominador? ¿Qué representa este numerador? ¿Y este otro?

- ¿Notan algún patrón en su cuadro? ¿Cómo podrían describirlos?

- ¿Qué observan acerca de la suma de las fracciones?

- ¿Qué observan acerca de la suma de los decimales?

▦ ¿Cómo pueden usar F⊃D para comparar fracciones y decimales?

▦ ¿Cómo pueden usar ÷ para comparar fracciones y decimales?

▦ ¿Usarían INT÷ para comparar fracciones y decimales? ¿Por qué?

Cómo analizar los datos y sacar conclusiones

Después de que los alumnos hayan reunido sus datos, pídales que analicen los resultados. Haga las preguntas siguientes:

- ¿La información de su cuadro cómo les ayuda a explicar los resultados de **Sumas con los cubos con números**?

- ¿Cuál fue la suma de sus probabilidades fraccionales? ¿Qué sumas obtuvieron los demás grupos? ¿Cómo pueden explicar eso?

- ¿Cuál fue la suma de sus probabilidades que mostraba la calculadora en forma decimal? ¿Cómo explican la diferencia?

- Hagan un informe de sus observaciones. ¿Cómo pueden explicar sus observaciones?

▦ Cuando transforman una fracción en un decimal y luego vuelven a fracción, a veces pueden usar la tecla F⊃D repetidamente y en otras ocasiones tienen que emplear la tecla x⊂y. ¿Por qué creen que ocurre esto?

▦ ¿Cómo pueden explicar lo que hace la calculadora cuando transforma fracciones en decimales?

Cómo continuar la investigación

Que los alumnos usen otros dados poliédricos y predigan cada resultado en el espacio muestreo. Que hagan una investigación similar con las representaciones fraccionales y decimales de la probabilidad de cada resultado posible.

Name:

Analyzing Number Cube Sums
Recording Sheet

Collecting and Organizing Data

Ways to Get Each Sum	Possible Sums	Fraction	Decimal
_____	2	_____	_____
_____	3	_____	_____
_____	4	_____	_____
_____	5	_____	_____
_____	6	_____	_____
_____	7	_____	_____
_____	8	_____	_____
_____	9	_____	_____
_____	10	_____	_____
_____	11	_____	_____
_____	12	_____	_____

Analyzing Data and Drawing Conclusions

Write about what your information shows you.

Nombre:

Cómo analizar las sumas con los cubos con números

Hoja de registro

Cómo reunir y organizar los datos

Formas de obtener cada suma	Sumas posibles	Fracción	Decimal
_____	2	_____	
_____	3	_____	
_____	4	_____	
_____	5	_____	
_____	6	_____	
_____	7	_____	
_____	8	_____	
_____	9	_____	
_____	10	_____	
_____	11	_____	
_____	12	_____	

Cómo analizar los datos y sacar conclusiones

Escriban acerca de lo que representa la información.

Picturing Probabilities of Number Cube Sums

<table>
<tr><td>

Math Concepts

- whole numbers
- fractions
- decimals
- ratio
- proportion
- multiplication
- angle measure
- circles
- perimeter
- area
- sample space
- probability

Materials

- Math Explorer™
- **Picturing Probabilities of Number Cube Sums** recording sheets
- data from **Analyzing Number Cube Sums**
- number cubes
- linking cubes
- protractors
- rulers
- large paper
- pencils

</td><td>

Overview

Students will use ideas of ratio and proportion to investigate various ways to make a circle graph. The graph will display the probabilities of the different sums that can be generated with two number cubes.

</td></tr>
</table>

Introduction

> The **Analyzing Number Cube Sums** activity on page 111 should be completed before beginning this activity.

1. Show some examples of circle graphs from newspapers and magazines and discuss them with students.

   Examples: Talk about what the sections represent, comparisons of sizes, the visual impressions they give that are different from bar graphs, etc.

2. Have students use the probabilities they found in **Analyzing Number Cube Sums** on page 111.

3. Have students represent the 11 different sums (2, 3, 4, 5, 6, 7, 8, 9, 10, 11, 12) in their data with different colors of linking cubes.

   Note: Linking cubes usually come in ten colors. You can place a sticky dot on one of the colors to distinguish it as the 11th color. Students will need 36 cubes to represent the 36 possible outcomes of tossing the two number cubes.

Representación de las probabilidades en las sumas con los cubos con números

Conceptos matemáticos

- números enteros
- fracciones
- decimales
- relación
- proporción
- multiplicación
- medida angular
- círculos
- perímetro
- superficie
- espacio muestreo
- probabilidad

Materiales

- Math Explorer™
- hojas de registro de **Representación de las probabilidades en las sumas con los cubos con números**
- datos de **Cómo analizar las sumas con los cubos con números**
- cubos con números
- cubos de enlace
- transportadores
- reglas
- papel grande
- lápices

Resumen

Los alumnos utilizarán las nociones de relación y proporción para investigar distintas formas de hacer un gráfico circular. El gráfico desplegará las probabilidades de las diferentes sumas que se pueden generar con dos cubos con números.

Introducción

> Antes de comenzar con esta actividad, se debe completar el ejercicio de **Cómo analizar las sumas con los cubos con números** en la página 111.

1. Muestre algunos ejemplos de gráficos circulares que aparezcan en periódicos y revistas, y analícelos con los alumnos.

    Ejemplo: analice lo que representan las secciones, las comparaciones de tamaños, las impresiones visuales que sean distintas a las de los gráficos de barras, etc.

2. Que los alumnos usen las probabilidades que encontraron en la sección **Cómo analizar las sumas con los cubos con números** de la página 111.

3. Que los alumnos representen las 11 sumas diferentes (2, 3, 4, 5, 6, 7, 8, 9, 10, 11, 12) en sus datos con los distintos colores de los cubos de enlace.

    Nota: los cubos de enlace generalmente vienen en diez colores. Pueden colocar una marca en uno de los colores para distinguirlo del 11º color. Los alumnos necesitarán 36 cubos para representar los 36 resultados posibles al lanzar los dos cubos con números.

Picturing Probabilities of
Number Cube Sums (continued)

Introduction (continued)

Example: Students will need one red cube to represent the one way to get a sum of 2, two blue cubes to represent the two ways to get a sum of 3, four green cubes to represent the four ways to get a sum of 4, etc.

4. Have students join all the linking cubes, keeping the same colors together, into a long bar. Then have them join the ends of the bar into a "circle" and place the "circle" on a large sheet of paper.

   Note: Students may have to lay the bar on a table allowing a few breaks in the bar to get it to form a circle. Or, the linking cubes with the hole through them can be strung on a string.

5. Have students sketch a "circle" around the inside of the linking cubes, estimate the center of the circle, and mark sections of the circle by drawing lines from the center to the circumference to show the number of cubes of each color. Discuss with students what these sections represent.

   Note: Students have made a sketch of a circle graph showing the probability of each sum.

6. Review with students the angle measure of a circle (360 degrees) and have them work together in small groups to predict the angle measures for each section in the circle graph. Then give the groups protractors so that they can evaluate their predictions and sketch their results on the recording sheet.

   Example:

Number Cube Sum	Probability	Angle Measure
2	1/36	10°
3	2/36	20°
4	3/36	30°

Representación de las probabilidades en las sumas con los cubos con números (continuación)

Introducción (continuación)

Ejemplo: los alumnos necesitarán un cubo rojo para representar la única forma de obtener una suma de 2, dos cubos azules para representar las dos formas de obtener una suma de 3, cuatro cubos verdes para representar las cuatro formas de obtener una suma de 4, etc.

4. Que los alumnos unan todos los cubos de enlace, manteniendo los colores iguales juntos y formando una fila. Después, que unan los extremos de la fila, formando un círculo y que coloquen el círculo en una hoja de papel grande.

5. Que los alumnos esbocen el círculo interior de la figura que forman los cubos de enlace, estimen el centro del círculo y marquen las secciones del mismo, dibujando segmentos que partan desde el centro de la circunferencia para mostrar el número de cubos de cada color. Analice con los alumnos lo que representan estas secciones

Nota: los alumnos hicieron un esbozo de un gráfico circular que muestra la probabilidad de cada suma.

6. Revise con los alumnos la medida angular de un círculo (360 grados) y pídales que trabajen en grupos pequeños para predecir las medidas angulares de cada sección del gráfico circular. Después, entregue a los grupos transportadores para que puedan evaluar sus predicciones y esbozar sus resultados en la hoja de registro.

Ejemplo:

Suma con los cubos con números	Probabilidad	Medida angular
2	$1/36$	$10°$
3	$2/36$	$20°$
4	$3/36$	$30°$

Picturing Probabilities of Number Cube Sums (continued)

Collecting and Organizing Data

While students are constructing their circle graphs and exploring with the calculator, ask questions such as:

- What does each linking cube represent?

- What part of the circle does each linking cube represent?

- How are you estimating the sizes of each section of the circle?

- How can you use linking cubes to predict the angle measure of each section?

- How can you use the fractions to help you predict the angle measure of each section?

- What should be true about the sum of the angle measures of all the sections?

▦ How can you use the calculator to help you find the angle measures?

▦ How can you use the calculator to help you determine if your predictions of angle measures are reasonable or not?

Analyzing Data and Drawing Conclusions

After students have made and evaluated their predictions, have them discuss their strategies as a whole group. Ask questions such as:

- How did you use the linking cubes to make your predictions of angle measures?

- How did you use the fractions to make your predictions of angle measures?

- What other strategies did you use?

- What strategies did you use to determine if your predictions were reasonable or not?

- What does the circle in the circle graph represent?

- What does each section in the circle graph represent?

- What advantages are there in presenting data in a circle graph?

- What disadvantages are there in presenting data in a circle graph?

▦ How did you use the calculator to help you in this problem?

▦ Were you able to stop using the calculator? Why or why not?

▦ Describe a similar type of problem where you might need the calculator more. **Example:** Suppose there had been 35 pieces of data instead of 36.

▦ When are calculators most useful?

▦ When are calculators not as useful?

Representación de las probabilidades en las sumas con los cubos con números (continuación)

Cómo reunir y organizar los datos

Mientras los alumnos construyen sus gráficos circulares y prueban con la calculadora, haga las preguntas siguientes:

- ¿Qué representa cada cubo de enlace?

- ¿Qué parte del círculo representa cada cubo de enlace?

- ¿Cómo están estimando el tamaño de cada sección del círculo?

- ¿Cómo pueden utilizar los cubos para predecir la medida angular de cada sección?

- ¿Cómo pueden usar las fracciones para predecir la medida angular de cada sección?

- ¿Qué debe cumplirse en la suma de las medidas angulares de todas las secciones?

▦ ¿Cómo pueden utilizar la calculadora para encontrar las medidas angulares?

▦ ¿Cómo pueden utilizar la calculadora para determinar si las predicciones de las medidas angulares son razonables o no?

Cómo analizar los datos y sacar conclusiones

Después de que los alumnos hagan y evalúen sus predicciones, pídales que analicen sus estrategias. Haga las preguntas siguientes:

- ¿Cómo usaron los cubos de enlace para hacer sus predicciones de las medidas angulares?

- ¿Cómo usaron las fracciones para hacer sus predicciones de las medidas angulares?

- ¿Qué otras estrategias utilizaron?

- ¿Qué representa el círculo en el gráfico circular?

- ¿Qué representa cada sección del gráfico circular?

- ¿Qué ventajas tiene la presentación de datos en un gráfico circular?

- ¿Qué desventajas tiene la presentación de datos en un gráfico circular?

▦ ¿Cómo utilizaron la calculadora en este problema?

▦ ¿Podían dejar de usar la calculadora? ¿Por qué?

▦ Describan un tipo similar de problema en que pudieran necesitar más la calculadora. **Ejemplo:** supongan que hubiera 35 datos en vez de 36.

▦ ¿Cuándo sirve más la calculadora?

▦ ¿Cuándo no sirve la calculadora?

Picturing Probabilities of Number Cube Sums (continued)

Continuing the Investigation

Have students:

- Estimate the areas of the sections using centimeter grid paper, record the areas as fractions of the whole circle's area, and compare the area fractions to the probabilities of the sums.

- Design a plan for making a circle graph from any set of given information.

Representación de las probabilidades en las sumas con los cubos con números (continuación)

Cómo continuar la investigación

Los alumnos deben:

- Estimar la superficie de las secciones, usando el papel cuadriculado en centímetros, registrar las superficies como fracciones de la superficie del círculo completo y comparar las fracciones de superficie con las probabilidades de las sumas.

- Diseñar un plan para hacer un gráfico circular a partir de un conjunto de datos determinado.

Name:

Picturing Probabilities of Number Cube Sums
Recording Sheet

Collecting and Organizing Data

Sketch your circle graph to represent the number cube sums.

Analyzing Data and Drawing Conclusions

Number Cube Sum	Probability	Angle Measure

We found the angle measure of each section by:

Nombre:

Representación de las probabilidades en las sumas con los cubos con números

Hoja de registro

Cómo reunir y organizar los datos

Dibujen su gráfico circular de manera que represente las sumas con los cubos con números.

Cómo analizar los datos y sacar conclusiones

Suma con los cubos con números	Probabilidad	Medida angular

Encontramos la medida angular de cada sección mediante el procedimiento siguiente:

An Average Lunch?

TI-108
MATH MATE
MATH
EXPLORER

Math Concepts

- whole numbers
- division
- addition
- mean
- multiplication

Materials

- TI-108, Math Mate™, Math Explorer™
- **An Average Lunch?** recording sheets
- pencils

Overview

Students will explore the meaning of the average (mean) value for a set of data.

Introduction

1. Have students read about Wayside School in *Sideways Stories from Wayside School* by Louis Sachar.

2. Present a scenario about managing the cafeteria at Wayside School. Tell students: It doesn't matter how much an individual lunch costs at Wayside School, just as long as the average price per lunch for the lunches sold is $1.85 each day.

3. Have students work in small groups to brainstorm different possible combinations of lunch prices that could satisfy this requirement. Have them record these combinations in the Data Set column on their recording sheets. Encourage students to use efficient mathematical notation.

   Example: 3($1.00) for three lunches costing $1.00 each.

4. Have students look for patterns in the sets of data they recorded in the table.

5. Have students write three summary statements on their recording sheets. These statements should describe characteristics that might be expected in a data set that has a given mean value.

Collecting and Organizing Data

While students are creating data sets for the given mean value, ask questions such as:

- What does it mean for the "average" lunch price to be $1.85?

 How can you use the calculator to find the mean of each data set?

¿Un almuerzo término medio?

Conceptos matemáticos

- números enteros
- suma
- multiplicación
- división
- media

Materiales

- TI-108, Math Mate™, Math Explorer™
- hojas de registro de ¿**Un almuerzo término medio?**
- lápices

Resume

Los alumnos examinarán el significado del valor medio (media) de un conjunto de datos.

Introducción

1. Que los alumnos lean acerca de Wayside School en *Sideways Stories from Wayside School*, de Louis Sachar.

2. Presenten una situación acerca de la administración de una cafetería en Wayside School. Indique a los alumnos lo siguiente: no importa cuánto cuesta un almuerzo individual en Wayside School, siempre que el precio medio por almuerzo sea $1.85 cada día.

3. Que los alumnos trabajen en grupos pequeños para reunir ideas sobre distintas combinaciones posibles de precios de almuerzos que pudieran cumplir esta condiciones. Pídales que registren estas combinaciones en la columna "Conjunto de datos" de la hoja de registro. Incentívelos para que usen una notación matemática eficiente.

   Ejemplo: 3($1.00) para tres almuerzos que cuestan $1.00 cada uno.

4. Que los alumnos busquen patrones en los conjuntos de datos que registraron en el cuadro.

5. Que los alumnos escriban tres afirmaciones concisas en las hojas de registro. Estas afirmaciones deben describir las características que cabría esperar en un conjunto de datos que tiene un valor medio determinado.

Cómo reunir y organizar los datos

Mientras los alumnos crean conjuntos de datos para el valor medio determinado, haga las preguntas siguientes:

- ¿Qué significa que el precio "medio" del almuerzo debe ser de $1.85?

 ¿Cómo pueden utilizar la calculadora para encontrar la media de cada conjunto de datos?

An Average Lunch? (continued)

Collecting and Organizing Data (continued)

- How many pieces of data do you want to have in your data set? Why? What is the least number possible? What is the greatest number possible?

- Do all the lunches have to cost the same amount? Can they?

- What happens when you sell a lunch for less than the mean value? What happens when you sell a lunch for more than the mean value?

▤ How can you use the calculator to help you find data sets that will produce a given mean?

▤ How can you use ⟦ and ⟧ to help you investigate this problem?

Analyzing Data and Drawing Conclusions

After students have investigated a variety of data sets that all have the same mean, have them discuss their observations as a whole group. Ask questions such as:

- What does a mean represent?

- What does the mean of a data set tell you about the number of pieces of data in a data set?

- What does the mean tell you about the values of individual pieces of data in the data set?

- What are some advantages of using the mean to describe a set of data?

- What are some disadvantages of using the mean to describe a set of data?

▤ How did you use the calculator to help you investigate this problem?

▤ How were you able to tell if the displayed values on your calculator were reasonable or not?

▤ How did you use ⟦ and ⟧ to help you investigate this problem?

Continuing the Investigation

Have students:

- Change the mean for the school lunches and investigate how their data sets need to change to produce the new given mean.

- Collect examples from newspapers and magazines illustrating uses of means and write short paragraphs evaluating these uses.

¿Un almuerzo término medio? (continuación)

Cómo reunir y organizar los datos (continuación)

- ¿Cuántos datos desean tener en su conjunto de datos? ¿Por qué? ¿Cuál es el menor número posible? ¿Cuál es el mayor número posible?

- ¿Todos los almuerzos deben costar lo mismo? ¿Pueden costar lo mismo?

- ¿Qué sucede cuando venden un almuerzo por un precio inferior al valor medio? ¿Y cuándo venden uno por un precio superior al valor medio?

⌨ ¿Cómo pueden utilizar la calculadora para encontrar conjuntos de datos que produzcan una media determinada?

⌨ ¿Cómo pueden usar las teclas ⟦ y ⟧ para investigar este problema?

Cómo analizar los datos y sacar conclusiones

Después de que los alumnos investiguen una variedad de conjuntos de datos que tengan la misma media, pídales que analicen sus observaciones. Haga las preguntas siguientes:

- ¿Qué representa una media?

- ¿Qué indica la media acerca de la cantidad de datos en un conjunto de datos?

- ¿Qué indica la media acerca de los valores de cada dato en el conjunto de datos?

- ¿Cuáles son algunas de las ventajas de usar la media para describir un conjunto de datos?

- ¿Cuáles son algunas de las desventajas de usar la media para describir un conjunto de datos?

⌨ ¿Cómo utilizaron la calculadora para investigar este problema?

⌨ ¿Cómo pudieron decidir si los valores desplegados en la calculadora eran razonables o no?

⌨ ¿Cómo utilizaron las teclas ⟦ y ⟧ para investigar este problema?

Cómo continuar la investigación

Los alumnos deben:

- Cambiar la media de los almuerzos e investigar cómo necesitan cambiar sus datos para producir la nueva media determinada.

- Reunir ejemplos a partir de periódicos y revistas que ilustren el uso de las medias y escribir párrafos breves que evalúen esos usos.

Name:

An Average Lunch?
Recording Sheet

Collecting and Organizing Data

For a mean value of:_____

Data Set	Total Cost	Number of Lunches

Analyzing Data and Drawing Conclusions

Characteristics we might expect in a data set with a mean of _____:

A.

B.

C.

Nombre:

¿Un almuerzo término medio?

Hoja de registro

Cómo reunir y organizar los datos

Para un valor medio de: _____

Conjunto de datos	Costo total	Número de almuerzos

Cómo analizar los datos y sacar conclusiones

Características que cabría esperar en un conjunto de datos con una media de: _____:

A.

B.

C.

Activity Content Index

ACTIVITY	PAGE NO.	PATTERNS	WHOLE NUMBERS	FRACTIONS	DECIMALS	INTEGERS	PLAVE VALUE
100 or Bust	2	x	Comparing				x
Random Reminders	6	x	x				
Recurring Remainders	10	x	x				
Remainder Rules	13	x	x				
Names for One-Half	18	x		x	x	x	
Patterns in Counting with Decimals	21	x			Comparing		x
Names for 100	24	x		x	x	x	
Area Patterns	28	Functions	Comparing				
Perimeter Patterns	33	Functions	Comparing				
The Mysterious Constant	38	Functions	Comparing				
CONSTANT-ly	42	Functions	Comparing				
"Power"ful Patterns	46	x	x				x
Patterns in Division	50	x	x	x	x		
Picturing Percents	55	x	x	x	x		x
How Owl Files	61	x	x	x			
Map It!	65	Functions	x				
Only Half There?	68	x		x			
No More Peas, Please!	71	x	x				
Do Centimeters Make Me Taller?	74	x		x	x		
What's My Ratio?	78	x		x	x		
Ratios in Regular Polygons	84	x		x	x		
Predicting π	88	x		x	x		
Spin Me Along	93	x	x	x	x		
Tiles in a Bag	98	x	x	x	x		
Does One = One?	102	x	x	x	x		
Number Cube Sums	107	x	x	x	x		
Analyzing Number Cube Sums	111	x	x	x	x		
Picturing Probabilities	114	x	x	x	x		
An Average Lunch?	119	x	x		Money		

Activity Content Index (continued)

OPERATIONS	ESTIMATION	RATIO	MEASUREMENT	GEOMETRY	PROBABILITY STATISTICS
Add. & Sub.					
Division					Graphing
Division					
Add., Sub., Multi., & Div.					
Add., Sub., Multi., & Div.					
Addition					
Add., Sub., Multi., & Div.					
Add. & Multi.	x	Similarity	Area	Similarity	
Add. & Multi.	x	Similarity	Perimeter	Similarity	
Add. & Multi.	x				
Add. & Multi.	x				
Multi., Div., & Exponents					
Division					
Add. & Div.		Proportion/Percent			
Addition			Linear		
Addition		x	Linear		
Division		x	Linear	Similarity	
Multiplication		x	Capacity/Volume		
Division		x	Linear		
Division		Similarity	Linear	Similarity	
Division		Proportion/Similarity	Perimeter	Similarity	
Division		Proportion/Similarity	Circumference	Circles	
			Area		x
					x
	Rounding				x
					x
					x
Multiplication		Proportion Area	Perimeter/Area/ Angle	Circles/Angles	Graphing
Add., Multi., & Division					Mean

Bibliography for Links to Literature

Baum, F.L., *The Wizard of Oz*. Tor Books, 1993.

Clement, R., *Counting on Frank*. Stevens Gareth Inc., 1991.

Heide, F., *The Shrinking of Treehorn*. Dell, 1979.

Milne, A.A., *Winnie the Pooh*. Dutton Children's Books, 1926.

Norton, M., *The Borrowers*. Harcourt Brace Jovanovich, 1981.

Peterson, J., *The Littles*. Scholastic Inc., 1967.

Peterson, J., *The Littles Go Exploring*. Scholastic Inc., 1978.

Peterson, J., *The Littles Take a Trip*. Scholastic Inc., 1968.

Peterson, J., *The Littles To the Rescue*. Scholastic Inc., 1968.

Sachar, L., *Sideways Stories from Wayside School*. Avon Books, 1985.

Stine, R.L., *Goosebumps — Monster Blood III*. Scholastic Inc., 1995.

Swift, J., *Gulliver's Travels*. Penquin Books, 1980.

Viorst, J., *Alexander and the Terrible, Horrible, No Good, Very Bad Day*. Atheneum, 1972.

Wilder, L.I., *Little House in the Big Woods*. Harper Trophy, 1971.

Wilder, L.I., *Little House on the Prairie*. Harper Trophy, 1971.